Surface Phenomena and Latexes in Waterborne Coatings and Printing Technology

Surface Phenomena and Latexes in Waterborne Coatings and Printing Technology

Edited by

Mahendra K. Sharma

Eastman Chemical Company
Kingsport, Tennessee

Plenum Press • New York and London

Library of Congress Cataloging-in-Publication Data

Surface phenomena and latexes in waterbone coatings and printing
technology / edited by Mahendra K. Sharma.
      p.   cm.
      "Proceedings of an International Symposium on Surface Phenomena
and Latexes in Waterbone Coatings and Printing Technology, held in
conjunction with the 23rd Annual Meeting of the Fine Particle
Society, July 13-17, 1992, in Las Vegas, Nevada"--T.p. verso.
    Includes bibliographical references and index.
    ISBN 0-306-45106-9
    1. Emulsion paint--Congresses. 2. Printing ink--Congresses.
I. Sharma, Mahendra K. II. International Symposium on Surface
Phenomena and Latexes in Waterbone Coatings and Printing Technology
(1992 : Las Vegas, Nev.) III. Fine Particle Society. Meeting (23rd
: 1992 : Las Vegas, Nev.)
    TP935.S834 1995
    667'.9--dc20   95-34648
                                                              CIP

Proceedings of an International Symposium on Surface Phenomena and Latexes in Water-Based
Coatings and Printing Technology, held in conjunction with the 23rd Annual Meeting of the Fine
Particle Society, July 13–17, 1992, in Las Vegas, Nevada

ISBN 0-306-45106-9

© 1995 Plenum Press, New York
A Division of Plenum Publishing Corporation
233 Spring Street, New York, N. Y. 10013

10 9 8 7 6 5 4 3 2 1

PREFACE

THE CURRENT STATE OF THE ART of waterborne polymers, paints, coatings, inks and printing processes is presented in this volume. This is the third volume in the series on waterborne coating and printing technology. It documents several invited papers and the proceedings of the International Symposium on Surface Phenomena and Latexes in Waterborne Coatings and Printing Technology sponsored by the Fine Particle Society (FPS). The FPS meeting was held in Las Vegas, Nevada, July 13-17, 1992. The volume deals with various basic and applied aspects of research on waterborne coating/printing technology. Major topics discussed involve waterborne polymers and polymer blends, pigment grinding, millbases, paint formulation, and characterization of coating films.

This edition includes sixteen selected papers related to recent developments in waterborne technology. These papers are divided in three broad categories: (1) Waterborne Polymers and Pigment Dispersions, (2) Latex Film, Wetting Phenomena and Printing Gloss, (3) Surfactants and Polymers in Aqueous Coating/Printing Systems.

This volume includes discussions of various waterborne polymers in coating/printing systems. The editors hope that this volume will serve its intended objective of reflecting the current understanding of formulation and process problems related to waterborne coatings, paints and inks. In addition, it will be a valuable reference source for both novices as well as experts in the field of waterborne technology. It will also help the readers to understand underlying surface phenomena and will enhance the reader's potential for solving critical formulation, evaluation and process problems.

I would like to convey my sincere thanks and appreciation to Professor M. S. El-Aasser, Director, Emulsion Polymer Institute, Lehigh University, for his assistance and support in organizing the symposium. I wish to convey my sincere thanks and appreciation to the members of the Fine Particle Society for their generous support. I would also like to express my thanks and appreciation to Ms. Patricia M. Vann and to the Editorial Staff of the Plenum Publishing Corporation for their continued interest in this project.

The editor is grateful to reviewers for their time and efforts in providing valuable comments and suggestions to improve the material presented in the manuscripts. I wish to convey my sincere thanks and appreciation to all authors and coauthors for their contributions, enthusiasm and patience. The views and conclusions expressed herein are those of the authors.

I wish to express my thanks to the appropriate management of the Eastman Chemical Company (ECC) for allowing me to participate in the organization of the symposium. My special thanks are due to Mr. J. C. Martin (ECC) for his cooperation and understanding during the tenure of editing this proceedings volume.

Finally, I would like to extend my sincere thanks to friends and colleagues for their assistance and encouragement throughout this project. Also I would like to acknowledge the assistance and cooperation of my wife, Rama, and extends the appreciation to my children (Amol and Anuj) for allowing me to spend many evenings and weekends working on this volume.

Mahendra K. Sharma
Research Laboratories
Eastman Chemical Company
Kingsport, TN 37662

CONTENTS

SURFACTANTS AND POLYMERS IN AQUEOUS COATING/PRINTING SYSTEMS

COLLOIDAL PIGMENT DISPERSION FOR

CORRUGATED BOARD

Dennis M. Wilson

S.C. Johnson Polymer
1525 Howe Street
Racine, WI 53403

ABSTRACT

The flexographic printing of corrugated board in the United States accounts for 95% of the corrugated market. There are three major segments within the market: high quality, medium quality and low cost. Emulsion synthesized polyelectrolytes are used as pigment dispersants and letdown vehicles in the later two segments. The design of polymeric dispersants for these applications requires the understanding of the colloidal character of the polyelectrolyte and the viscosity profile as functions of pH. A series of five polyelectrolytes were synthesized. These copolymers were characterized and formulated as dispersants and letdown vehicles in carbon black inks. It was found that the molecular weight and the T_g of the copolymers were important variables in the design for maximizing the jetness in the formulation of a corrugated ink. An attempt was made to correlate the printing ink properties and performance parameters with various properties of the copolymers usually used as a binder in the printing processes.

INTRODUCTION

The low molecular weight anionic surfactants have been utilized for many years in a wide range of applications such as pigment dispersions. These surfactants are classified by a variety of characteristics, among the more important are low solubility in water, the ability to form micelles at higher

Surface Phenomena and Latexes in Waterborne Coatings and Printing Technologies, Edited by M.K. Sharma, Plenum Press, New York, 1995

1

concentrations and to stabilize or emulsify insoluble materials. The effective surfactants have a balanced ratio of hydrophilic to hydrophobic moieties within certain broad ranges.

Polymeric anionic surfactants are a unique class of materials in which the size of the molecule and the ratio of hydrophilic to hydrophobic moieties can be varied at will by copolymerization. This paper describes the aqueous solution characteristics for a limited series of copolymers and terpolymers of ethyl acrylate (EA), methyl methacrylate (MMA) and methacrylic acid (MAA) synthesized by emulsion polymerization. These polyelectrolytes were studied for their utility in dispersing carbon black pigment and the making of inks for the flexographic printing of medium quality and lower cost corrugated stock, which is briefly described herein.

EXPERIMENTAL

The copolymers were prepared with commercial monomers. A typical copolymerization comprises sodium dodecyl sulfate as the emulsifier, the thermal decomposition of ammonium persulfate during a two hour semi-batch addition of monomers under inert atmosphere in a glass reaction flask; the reaction mixture held for one half hour then cooled to room temperature; typically the solids level was 40%. The molecular weight was regulated through the use of a mercaptan chain transfer agent.

The particle size measurements were made in aqueous solution using a quasi elastic light scattering (QELS) technique with a Brookhaven Instruments BI-90 AT. The viscosity measurements were made with Hercules Viscometer DV-10 with bob size E, 4400 maximum rpm, a ramp time of 20 seconds at 24°C, at 20% polymer solids; and a Brookfield DV-II Viscometer at 25°C. The glass transition temperatures were obtained using Perkin-Elmer DSC-7. The carbon black pigment dispersions were run in a 24 hour ball mill using Elftex 8 pigment.

BACKGROUND

The stages of incorporating a pigment into a vehicle requires several attributes of the copolymer during the wetting, grinding and dispersing[5]. Wetting can be described as the displacement of air from the recesses of the pigment agglomerates by a fluid, in this case an aqueous solution. The velocity of the vehicle entering the pigment particle interstices has been described by:

$$v = \frac{\gamma \cos \theta 3}{4L} \frac{r}{\acute{\eta}}$$

where v is the velocity, r and L are the radius and length of the interstitial spaces, γ is the surface tension of the grinding vehicle, θ is the pigment/vehicle contact angle and $\acute{\eta}$ is the viscosity of the vehicle. Thus, to practically maximize the pigment wetting, the surface tension of the solid should greater than that of the vehicle, the contact angle of the vehicle (large cos θ) and the vehicle viscosity should be low. During the grinding, new surfaces of the pigment are then wet as the agglomerates are broken up. This requires the transfer of the mechanical energy into the pigment slurry so that the entire surface of each particle is exposed for wetting. The viscosity of the slurry under shear must be maintained within a broad range in order to optimize the wetting process and allow for the maximum mechanical energy transfer.

The surface tension of the vehicle and its viscosity are the two attributes that can be modified. The polyelectrolytes described here were synthesized via emulsion polymerization at an acidic pH but are utilized for pigment dispersion at a basic pH. During the preparation, a low molecular weight surfactant was used to stabilize the emulsion colloid. Upon neutralization of the colloidal acids, the surfactants are redispersed in the solution. Most surfactants used in emulsion polymerization have lower surface tensions than the neutralized polyelectrolytes thus contribute during the pigment wetting process.

During the dispersing phase the wetted primary pigment particles that are stabilized are permanently separated in the vehicle. The particles at this stage will have various levels of adsorbed polyelectrolyte thus imparting an electrical charge on the surface of the particle. In this paper, the negatively charged copolymer is the lower hydrated adsorbant ,with the much more highly hydrated cation, ammonium hydroxide, closely associated with the particle. The more dense cloud of oppositely charged counter ions surrounding the anionically charged particle surface is commonly referred to as the ionic double layer. An electrostatic potential created by the nonuniform distribution of ions within the solution is highest near the surface of the particle and decreases with distance from the particle. Repulsion occurs between similarly charged clouds which is the essence of the stabilizing effect of the adsorbed polyelectrolyte. The electrostatic repulsive forces within a system are shown to increase the size of the double layer. A measure of the stabilizing effect of the polyelectrolyte is the particle size of pigment which is directly reflected in the color development of the grind. In the case of carbon black dispersions for corrugated, the jetness or blackness of the ink is indicative of the particle size. The optimum is the maximum color (i.e.: finest particle size) from the least amount of pigment.

The control of the viscosity of the vehicle and grind stability are the major contributions of the polyelectrolyte. Earlier studies[1-4] have shown that copolymers can be made by emulsion polymerization with a wide range of hydrophobicities and T_g's. The results of these studies indicated that the hydrophilic monomers were crucial to obtaining a resinous copolymer that was non-colloidal upon neutralization. Although the work was primarily focused on the alkali-thickening of latexes, the figure from Verbrugge[1] (Figure 1) shows MMA and EA containing copolymers in conjunction with molar ratios of MAA greater than one to five would have an open or rod-like configuration at high degrees of neutralization. All of this

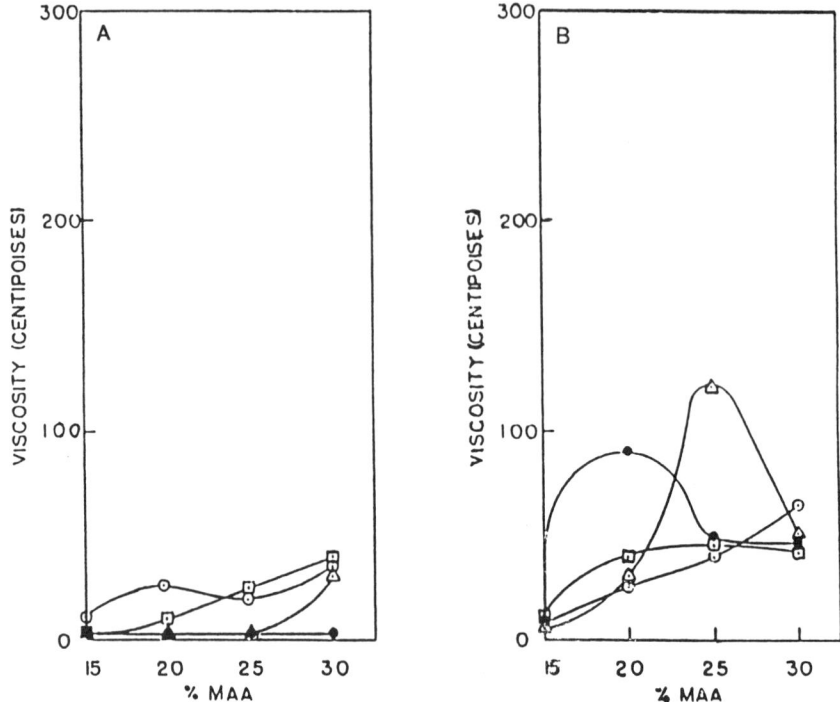

Figure 1. Representative Plots of Plateau Viscosities
Versus % MAA.
A: S/EA/MAA (⊙) 60% EA; (⊡) 40% EA; (△) 20% BA;
(●) 0% EA.; B: MMA/BA/MAA (⊙) 60% BA; (⊡) 40% BA;
(△)20% BA;(●) 0% BA. (Reproduced with Permission
from Reference 1).

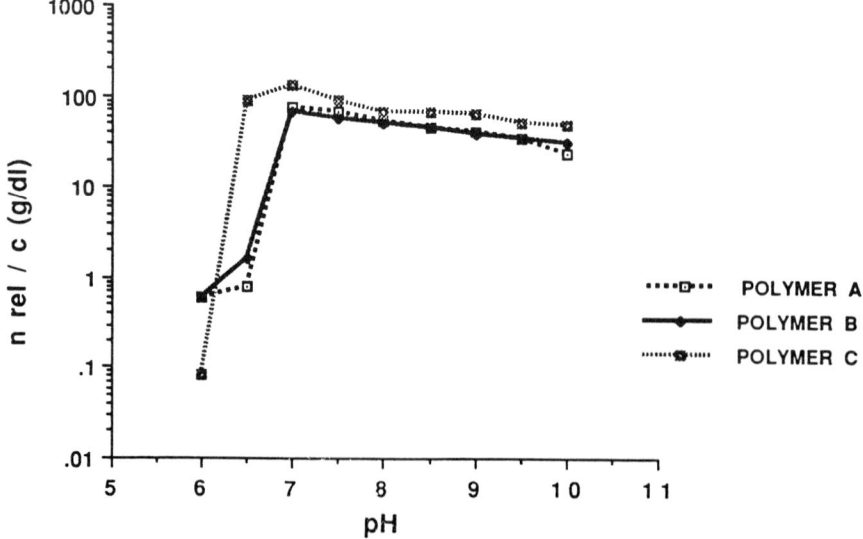

Figure 2. Comparison of the pH and Reduced Viscosity Profiles
for Polymers A, B and C.

4

previous work was done with copolymers having molecular weights (M_w) greater than 300,000. As the degree of neutralization increased, the configuration changed from a compact colloid, to a swollen colloidal particle followed by complete solublization. The elucidation of the configurational change regions for the copolymers and the resulting rheology is important to understanding the utility of polyelectrolytes in pigment dispersion.

The preparation of an ink is usually carried out in two stages, a grinding stage and letdown stage. The previous discussion involved the description of the conditions leading to the optimization of the pigment dispersion. The later stage involves reducing or letting down the grind with additional vehicle which may or may not be the same vehicle used in the grind. This additional vehicle is used to optimize the performance and application properties of the final ink.

RESULTS

The following five polyelectrolytes were synthesized to obtain three molecular weight regimes for this study. All had molar ratios of co-monomer to MAA moieties of four to one. The ratio was selected by using the information described in previous work[1-4] showing the lowest level of acid necessary to obtain a fully rod-like configuration upon neutralization. The molecular weights and glass transition temperatures of the copolymers are listed in Table I.

The glass transition temperatures are onset temperatures; they were modified by changing the ratio of EA to MMA in the terpolymers. The molecular weights are weight averages with polydispersities of 5-7.

Table I. Molecular Weight and Glass Transition
Temperature of Several Copolymers.

COPOLYMER	T_g	MOLECULAR WEIGHT
A	36	25,000
B	16	29,000
C	21	52,000
D	39	66,000
E	30	100,000

VISCOSITY PROFILE

All the copolymers were neutralized with ammonium hydroxide to the specified pH and allowed to equilibrate for 24 hrs. at 25°C prior to the Brookfield viscometer measurement. Figure 2 compares the reduced viscosity profiles for the two low molecular weight copolymers, A and B, and the intermediate molecular weight copolymer C. Comparing the low molecular weights, the glass transition temperature, T_g, difference showed only a very nominal change in the viscosity profiles particularly in the peak viscosity region. Polymer C has a relatively low T_g similar to Polymer B and the viscosity profile for C reflects the molecular weight increase over B. It is important to note that all three viscosities do go through a maximum at a pH just lower than 7.

The substantial increase in the reduced viscosity profile for Polymer D shown in Figure 3 clearly indicates the effect of higher T_g compared to Polymer C. Although copolymer D is slightly higher in molecular weight than C, the dominant variable is the T_g. Evidence for this is the broadening of the peak viscosity region even up to the pH levels of 8 - 8.5. This was supported by the work in references 3 and 4 at much higher molecular weights.

Figure 4 compares the viscosity profile of the highest molecular weight copolymer, E, to Polymer D. The peak viscosity is very evident about pH 7 but drops quickly as the pH is increase between pH 7 and 8. The higher T_g Polymer D maintains a higher viscosity above pH 7.5. The dominance of T_g over molecular weight to sustain a higher viscosity will be supported further by the particle size information in the next discussion. All of the pH/viscosity profiles indicate a relatively flat region exists above pH 8.0 which is important for the robustness of the ink formulation work described later in this paper.

PARTICLE SIZE AND SIZE DISTRIBUTION

The particle size distributions were measured by QELS to compare the transition regions for each copolymer as the pH is increased. The limitation of QELS is that it emphasizes larger light scattering particles where as electron microscopy emphasizes the smaller particles. As a result the ordinate for the following graphs of the distributions is Relative Intensity. Information can be obtained about the positioning of the peaks and the ratio of the peak areas (relative intensity, volume and number of light scattering particles) but not about the breadth of the distributions or peak widths. The peak width information can be made relevant only by running the experiments for orders of magnitude longer time frames.

Both of the lower molecular weight copolymers, A and B, had the same particle size distribution response with increasing pH. As a result only the distributions for Polymer A are shown in Figure 5. At a pH of 6.4 the mean particle diameter was 121 nM

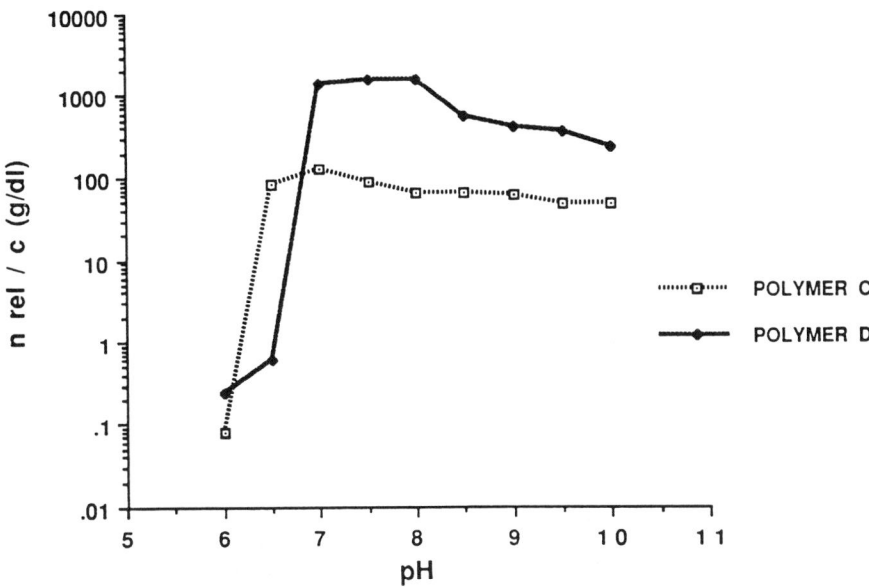

Figure 3. Comparison of the pH and Reduced Viscosity Profiles for Polymers C and D.

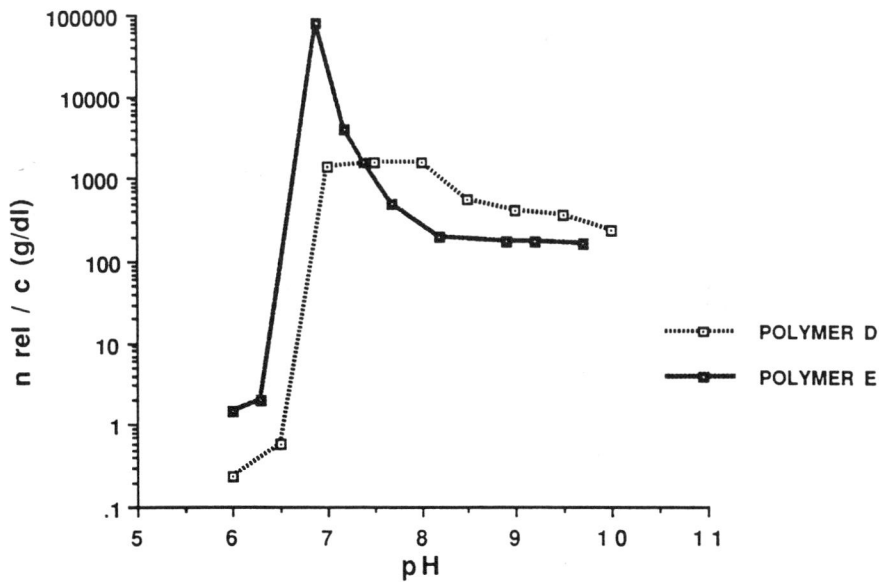

Figure 4. Comparison of the pH and Reduced Viscosity Profiles for Polymers D and E.

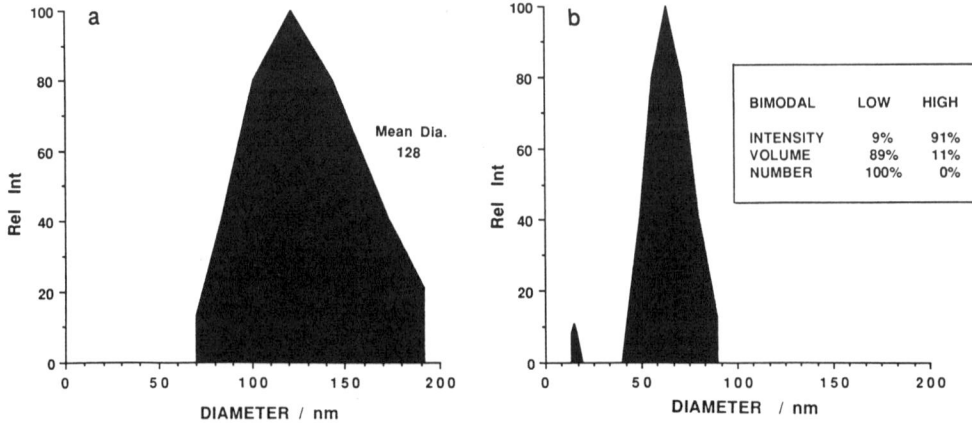

Figure 5. Comparison of Particle Size Distributions for Polymer A at Different pH. (a): pH = 6.4; (b): pH = 6.8.

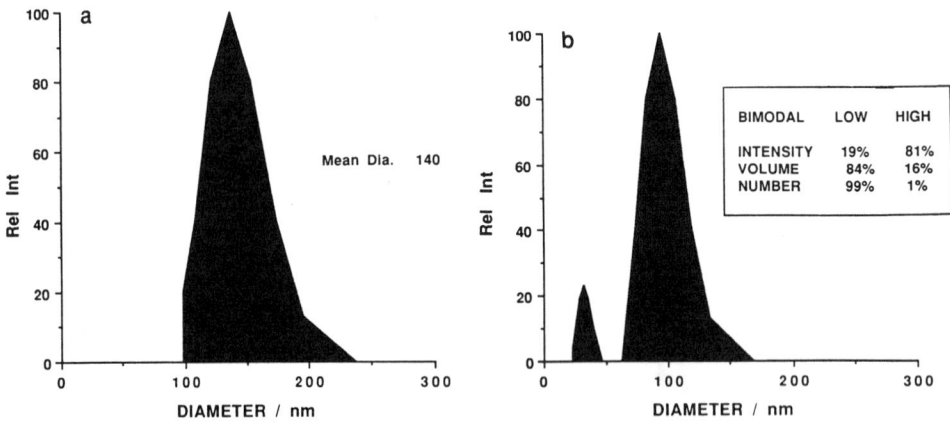

Figure 6. Comparison of Particle Size Distributions for Polymer C at Different pH. (a): pH = 6.4; (b): pH = 6.8.

and changed to a bimodal distribution upon further neutralization to pH 6.8. The relative intensity indicates that the distribution of particles is dominated by higher peak with a mean of 63 nM over the lower with a mean of 15 nM. In actuality the volume and number ratios indicate just the opposite is more realistic, which is a consequence of the type of measurement as stated above. With both of the mean particle sizes lower than the mean at pH 6.4, the copolymer is being solubilized away from the particle and not swelling. The dissolved polymer causes the observed increase in the viscosity as a function of pH. The particle sizes were too small to measure at a pH of 7.2.

Polymer C showed a very similar trend (see Figure 6) with the distribution centered around a mean diameter of 137 nM at pH 6.4. The similar particle size shift was observed as the pH was raised to 6.8 with mean diameters 32 nM and 94 nM. These bimodal distributions were weighted heavily for volume and number ratios toward the lower distribution. Again the copolymer was dissolving away rather than remaining with the particle in a colloidal form and swelling. Measurements at higher pH were again not possible.

Polymer D was quite different than the three polyelectrolytes discussed previously. At the lower pH of 6.5 a mono-modal particle size was observed with a mean diameter of 146 nM as shown in Figure 7. The distribution at pH 7.0 was quite multimodal with mean diameters at 12, 54, and 272 nM (not shown). At pH 7.5 the distribution was still multimodal (Figure 7b) with some swelling still observed prior to total dissolution. The mean particle diameters were observed at 26, 111, and 685 nM. The ratios for the three peaks were again heavily weighted for volume and number ratios to the lower two indicating the continued dissolution of the particles. The observed colloidal character surpassing pH 7.5 is consistent with the broadened viscosity profile observed for this copolymer. The combination of higher T_g and higher molecular weight appears to impede the ability of the ionized copolymer to dissociate with the colloid compared to the lower T_g analog, Polymer C.

The highest molecular weight copolymer, E, showed significant swelling prior to dissolution when the particle size distributions are compared (Figure 8) at pH's 6.5 and 6.9. At the lower pH the mean diameter 144 nM is very similar to the other copolymers. The bimodal distribution at the higher pH with mean diameters of 54 and 770 nM have an equivalent volume ratio. The number ratio again favors the lower distribution. The total dissolution with increasing pH above 6.9 is consistent with the observed rapid rise and fall of the pH/viscosity profile through this same region for this high molecular weight polyelectrolyte.

SHEAR RATE - VISCOSITY PROFILE

The pigment grinding process involves shear rates from 1000 sec^{-1} to 100,000 sec^{-1}. The Brookfield viscosities reflect shear rates from 0.1 sec^{-1} up to 10 sec^{-1} which give limited information about the flow and leveling of the copolymer in an

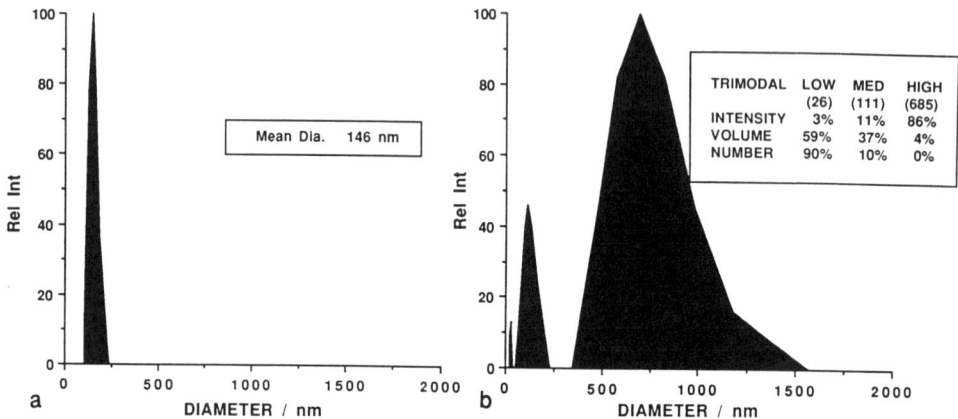

Figure 7. Comparison of Particle Size Distributions for Polymer D at Different pH. (a): pH = 6.5; (b): pH = 7.5.

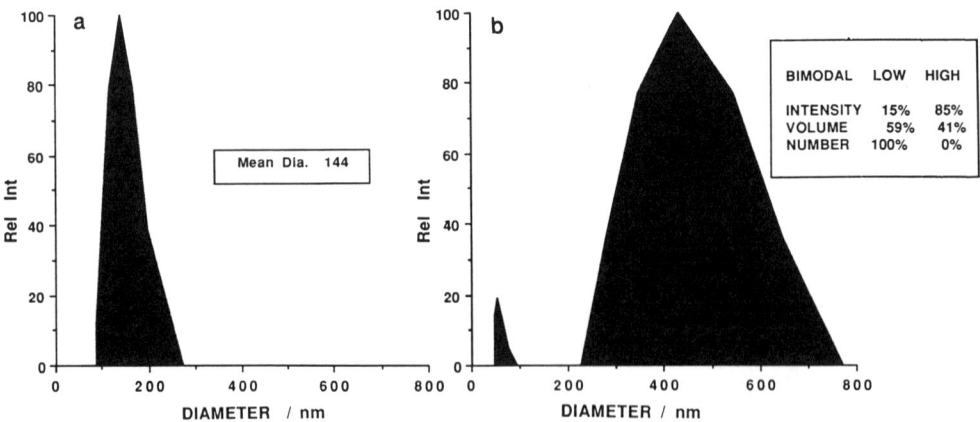

Figure 8. Comparison of Particle Size Distributions for Polymer E at Different pH. (a): pH = 6.4; (b): pH = 6.9.

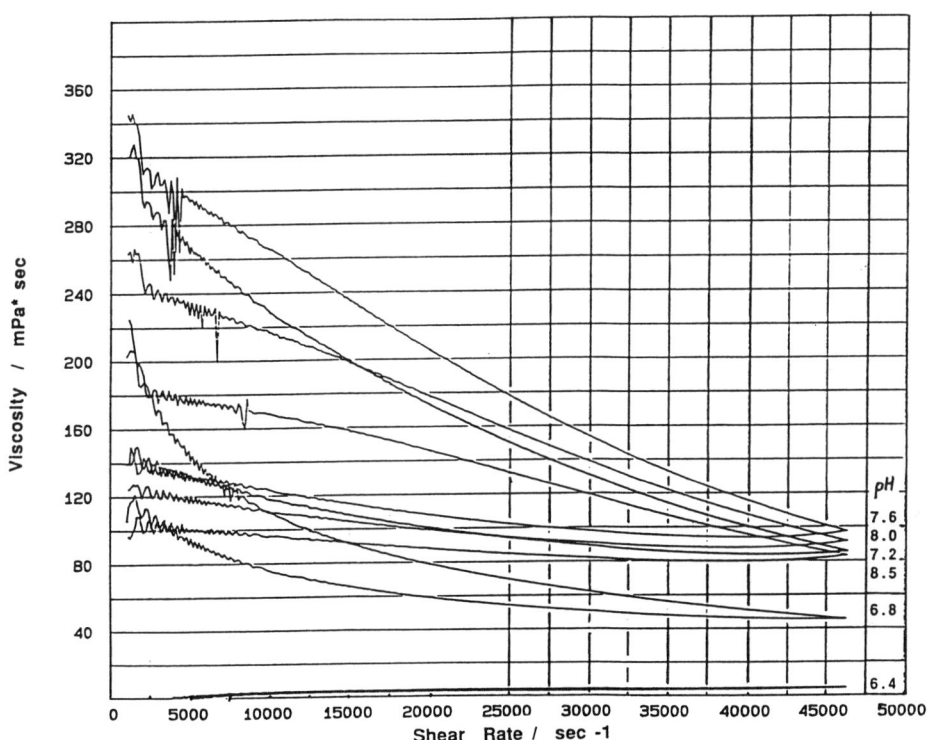

Figure 9. Comparison of the Shear Rate and Viscosity Profiles as a Function of pH for Polymer A.

ink upon application. Information was obtained about the viscosity response of two of the copolymers (A and C) to increasing and decreasing shear rates as a function of the pH of the 20 % solids solutions. The shear rate range was increased from 1000 to 45,000 sec^{-1}. Figures 9 and 10 show the viscosity response for each of the copolymers A and C respectively. As the pH increased the thixotropy of these solutions became more evident and the hysteresis in the viscosity became larger. Both polymers A and B, as reported above, were not fully through their transition region until pH 7. This data indicates that the transition extends above pH 7.5 as the viscosities at low shear rate increase up to that point. The final viscosities at maximum shear rate for both copolymers converge to regions of plus or minus 20 cps above pH 7.5.

The hysteresis areas for the two copolymers at each solution pH were plotted in Figure 11. For both copolymers the areas increase to a maximum about pH 7.6 then start to decrease. An explanation for these results is that at low pH the copolymer is still in colloidal form and the interactions between colloids is small, thus the very low initial viscosity response to increasing shear rate and quick recovery to decreasing shear rate. As the copolymer undergoes the colloid to solvated polyelectrolyte transition the polymer- polymer interactions in solution increase. This results in the observed increased viscosity response to increasing shear rate and the slower recovery to decreasing shear rate. At the high pH the high ionic strength causes repulsion between the polyelectrolytes and results in lower initial viscosity response to increasing shear rate. The recovery appears to be the same at pH 8.5. Considering that the polymer-polymer interactions dominate the viscosity response to shear rate, the increase in the hysteresis area with increasing molecular weight (comparing Polymers A and C) was to be expected. Similar increases in viscosity hysteresis would be expected from higher molecular weight copolymers.

CORRUGATED INKS

Based on the data generated above, these copolymers could be utilized at pH 7.0 or greater if the residual colloidal character were the only consideration. The viscosity profiles indicate that a flatter response to small fluctuations in pH exits above pH 8.0. Finally, the shear rate profiles indicate that possibly there may be an advantage to stabilizing the polyelectrolyte above pH 8. The following tables (Tables II and III) show the ink formulation work done with the five copolymers. Each of the copolymers was neutralized to a pH of 8.2 with ammonium hydroxide. The carbon black grind formulas were adjusted to accommodate the differences in the viscosities of the copolymers. Even with this adjustment the grinds with Polymers C and D slightly high in viscosity.

Table III shows examples of the letdown formulas for each of the five pigment grinds reduced with the same copolymer. The matrix of formulas for each of the grinds letdown with the other copolymers was not included in this paper.

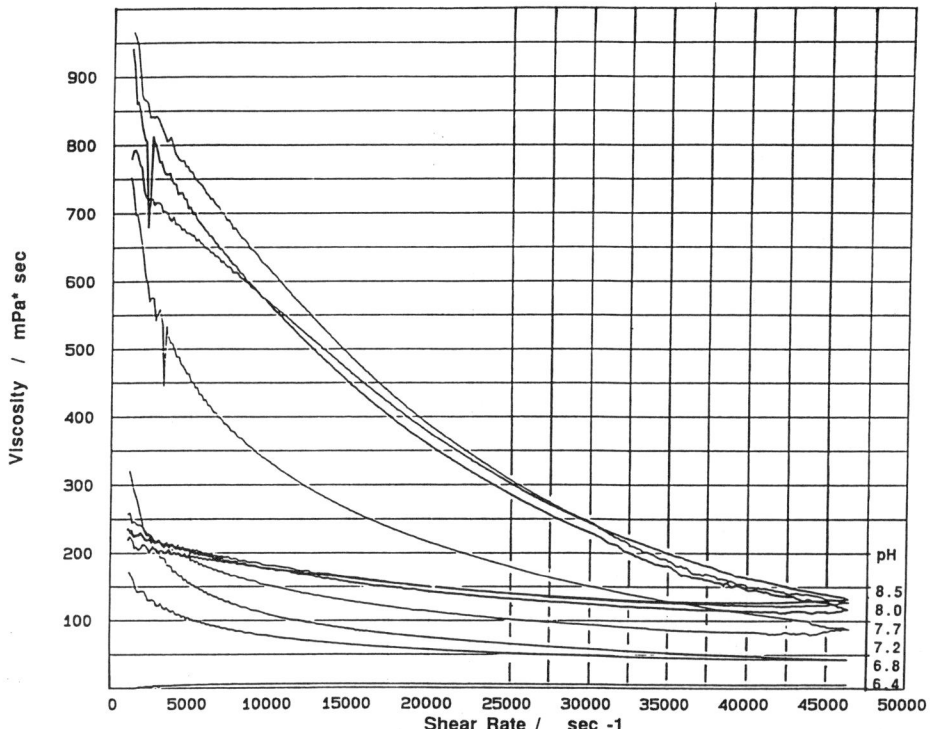

Figure 10. Comparison of the Shear Rate and Viscosity Profiles as a Function of pH for Polymer C.

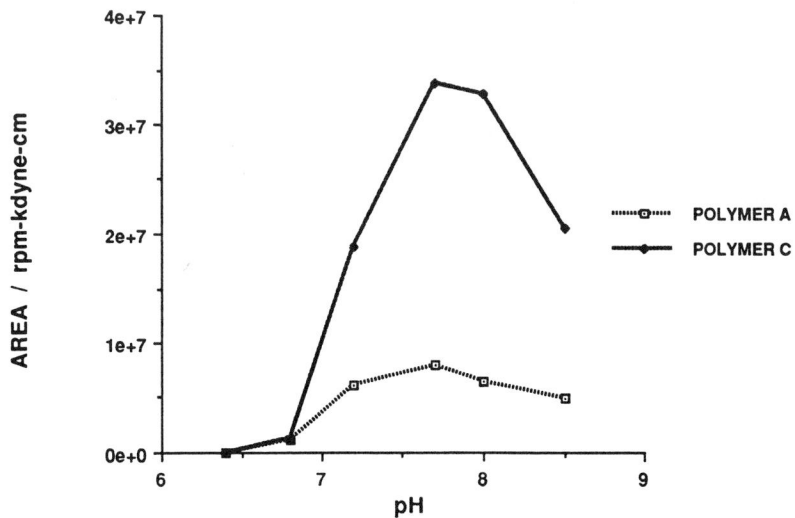

Figure 11. Comparison of the Hysteresis Areas as a Function of pH for Polymers A and C.

Table II. Carbon Black Grinds

POLYMERS	A	B	C	D	E
Vehicle @ 20% NV	30.0	40.0	30.0	30.0	23.0
Elftex 8*	25.0	16.0	25.0	25.0	15.0
DF-75**	0.5	0.5	0.5	0.5	0.5
APF-411***	1.0	1.0	1.0	1.0	1.0
Water	43.5	42.5	43.5	43.5	60.5
pH	8.1	8.1	8.1	8.1	8.1
Viscosity, cps	520.0	40.0	5100.0	4950.0	240.0

* Carbon black - Cabot Corp.
** Defoamer- Air Products Corp.
*** Micronized wax - Micro Powders Corp.

All of the inks were formulated to the same pigment level (12 %) and the pigment to binder ratio allowed to float to again accommodate the differences in copolymer viscosity. The viscosities of the inks were adjusted to 22 seconds plus or minus 1 second #2 Zahn cup.

Table III. Black Inks

POLYMERS	A	B	C	D	E
Base Grind	48.0	75.0	48.0	48.0	80.0
Polymer A @ 20% NV	40.0	-	-	-	-
Polymer B @ 20% NV	-	25.0	-	-	-
Polymer C @ 20% NV	-	-	25.0	-	-
Polymer D @ 25% NV	-	-	-	27.0	-
Polymer E @ 20% NV	-	-	-	-	5.0
APF-411	0.5	1.0	0.5	0.5	0.8
DF-75	0.2	0.2	0.2	0.2	0.4
Water	11.3	-	26.3	24.3	13.8
Viscosity #2 Zahn (Sec.)	21.0	21.0	22.0	21.0	23.0
% Pigment	12.0	12.0	12.0	12.0	12.0
% Polymer Solids	10.9	11.0	7.9	10.4	4.7
pH	8.1	8.3	8.1	8.1	8.8

Table IV shows the evaluation results after the inks were drawn down on standardized corrugated stock with #6 wire wound rod and with a 120 line anilox roll. Comparing the two methods of application gives an indication if there will be any transfer problems with the formulation. The jetness of the ink on the stock is the measure of how well the pigment was ground and stabilized at that particle size. The finer the pigment is ground the more color or jetness shows up in the ink, therefore the better the vehicle for the application. Typically a densitometer is used to measure the jetness, but corrugated stock is quite irregular. The readings are not found to be totally reproducible thus a qualitative scale with side-by-side drawdowns must also be used for the evaluations.

Table IV. Evaluation results of the dried ink film on corrugated surface.

L	POLYMERS	BASE GRINDS				
E		A	B	C	D	E
T	A	5	5	5	4	X
D	B	5	5	5	4	X
O	C	5	5	5	4	X
W	D	5	5	5	4	X
N						
S	E	3	3	3	3	1

Relative scale: 1 = poor, 5 = good.

The initial evaluation results of the inks described in Table III are shown in Table IV across the diagonal from top left to bottom right. It is apparent that the high molecular weight copolymer E is not an effective grind vehicle in this type of formulation. The further evaluation using Polymer E in the grind portion of the formulas was eliminated. Polymer E used in the letdown portion of the formulas appears to significantly reduce the jetness of the ink with all of the base grinds of the lower molecular weight copolymers. Using Polymer D as the grinding vehicle did show a slight decrease in the jetness of the ink but using it as the letdown vehicle for the pigment grinds of the three lower molecular weight copolymers showed no deleterious effect. The matrix of results for the three copolymers A, B and C indicated that they were very good grinding vehicles. The jetness was also maximized when were used as letdown vehicles interchangeably for the lower molecular grinds.

It appears to be a limit to the molecular weight and T_g of the copolymers used in the base grind of the ink formulas. Both of these variables directly influence the viscosity of the vehicle and thereby the amount of vehicle used in the base grind. Utilizing higher levels of copolymer in the vehicle would impede the transfer of mechanical energy during the grinding process. Further, the level of polyelectrolyte in the grind during the dispersing phase determines the level of stabilizer available to adsorb onto the pigment particles. Lower levels of a stabilizer may increase the pigment particle size thus decrease the jetness observed in the finished ink.

CONCLUSIONS

Significant differences were observed among the five polyelectrolytes examined in this study for their effectiveness in the dispersing and letdown portions of the corrugated ink formulas. The lower molecular weight polyelectrolytes A, B, and C are all very effective in either portion of the ink making process. The higher T_g of Polymer D appears to limit the ability of this polyelectrolyte to fully solubilize from the colloid which impedes the stabilizing effect during the dispersing process. This does not appear to be a detriment when Polymer D is used as the letdown. The higher vehicle viscosity of Polymer E severely limits its utility as a dispersant and as a letdown in maximizing the jetness in a flexographic corrugated ink formulation. The economic impact or optimization of copolymer usage were not taken into consideration since they were beyond the scope of this paper.

ACKNOWLEDGMENT

The author wholeheartedly thanks Mr. Daniel Demant, Mr. Thomas Czarnecki, Mrs. Florence Todd and especially Mr. James Cremer for without whose help this paper would not have been possible.

REFERENCES

1. Verbrugge, C.J., "Mechanism of Alkali Thickening of Acid-Containing Emulsion Polymers. I. Examination of Latexes by Means of Viscosity", J. Appl. Pol. Sci., **14**, 897, (1970).

2. Verbrugge, C.J., "Mechanism of Alkali Thickening of Acid-Containing Emulsion Polymers. Il. Examination of Latexes with the Light Microscope", J. Appl. Pol. Sci., **14**, 911 , (1970).

3. Nishda, S., Kast, H., El-Aasser, M.S., Vanderhoff, J.W., "Alkali Swelling Behavior of Carboxylated Latexes 1.", EPI, Lehigh Univ., (1979).

4. Nishda, S., Kast, H., El-Aasser, M.S., Vanderhoff, J.W., "Alkali Swelling Behavior of Carboxylated Latexes ll.", EPI, Lehigh Univ., (1979).

5. Shinoda, K., Tamamushi, B., Nakagawa, T., Isemura, T., Colloidal Surfactants, Academic Press, (1983).

EFFECTIVENESS OF ACRYLIC POLYMERS IN WATER-BASED

PIGMENT DISPERSIONS - A TAGUCHI APPROACH

Steven Y. Chan

Air Products and Chemicals, Inc.
Polymer Chemicals Division
Allentown, PA 18195

ABSTRACT

Pigment dispersion is a critical step in manufacturing printing inks. Styrene/acrylic copolymers and acrylic polymers are the common water based resins used in the graphic arts industry to disperse pigments. While surfactants may aid in pigment dispersions, complete surfactant grind pigment dispersions are only used in special applications. In this paper, three styrene/acrylic copolymers and two acrylic polymers with molecular weight (M_n) ranging from 1,700 to 4,800 and 25,000 to 42,000 respectively were evaluated in pigment grinding. Carbon black was chosen as the pigment because black is one of the colors used in largest volume in printing inks. Original layout of the experiments will require a total of 4374 (6^1 x 3^6) trials to cover all the combinations that it intends to study. Instead, the Taguchi's Orthogonal Array experimental design was employed. An L18 (2^1 x 3^7) design was chosen and modified as L18 (6^1 x 3^6). This technique enables us to study one factor at six levels and six factors at three levels each by performing eighteen experiments only. It was found that styrene/acrylic copolymers provide high jettness and gloss for the black grind. Acrylic polymers develop higher color strength than the copolymers. The pigment to binder

Surface Phenomena and Latexes in Waterborne Coatings and Printing
Technologies, Edited by M.K. Sharma, Plenum Press, New York, 1995

19

ratio (range from 5/1 to 3/1) and grinding time (10 to 30 minutes) had no effect on the quality of the pigment dispersions. The resins studied are all effective within these constraints. The defoamer and surfactant used in this study showed an adverse effect on the dispersions.

INTRODUCTION

Pigment dispersion is a critical step in manufacturing printing inks. Generally, dry pigments or presscakes are ground to very fine particle size and dispersed in a liquid form stabilized by a polymer. Then, ink manufacturers will take these stable pigment dispersions and "letdown" with their "ink vehicle" to "finished inks" by simply blending them together. Additional water or solvent may be added to adjust the final viscosity. Since most pigments are far more expensive than the dispersing resins, the effectiveness of the resin in developing desired properties is very important. One would like to develop as much color strength and gloss as possible from the pigment in a stable fluid system.

Solvent based printing inks have dominated the graphic arts market for years. However, contending with ever growing air quality rules at state, federal and international levels, especially with the Clean Air Act Amendments in 1990, many converters are being forced to change to water based systems. Acrylic and styrene/acrylic polymers are commonly used in water based inks. In this paper, three styrene/acrylic copolymers, two acrylic polymers and their combinations, all water based, are used to study the pigment dispersing efficiency. They vary in their molecular weight and therefore, their solution viscosities differ accordingly.

Black pigment is one of the largest volume colors used in printing inks. Moreover, commercial carbon black pigments vary in particle size, surface area, and structure (referred as the morphology of the primary aggregate) all of which affect the dispersion. However, the quality of the carbon black dispersion can be easily defined by measuring the jettness, gloss, color strength and fineness of grind. All these factors make carbon black the ideal choice of pigment for this study.

The process of pigment dispersion is normally very time consuming, required high energy/shear and special equipment. Even in the laboratory, it typically requires more than three hours to prepare one dispersion with the shot mill, not including the time in evaluating the properties of the dispersion. The formulations and processing factors, for example, mix ratio of polymers, pigment/binder ratio, grinding time, level of defoamer and surfactant, just to name a few, all play an important role in the pigment dispersion. If the conventional approach (one variable/one step at a time) is employed, formulators may not be able to evaluate all possible combinations due to prohibitively large number of experiments. This will prevent one from detecting the effects resulted from the interaction of a dynamic system with various controlling factors. Instead, the Taguchi's experimental design[1-3] was used in this study.

Taguchi's Orthogonal Array is a statistical design based on the mathematical concept of orthogonal arrays. The number of experiments is reduced tremendously by trading the error of degrees of freedom for the assessment of controllable factors or interactions. The properties and performance of the pigment dispersions are measured. Utilizing the analysis of variance technique with the computer, a full profile of properties/factors relationship was developed. The strategy of the Taguchi experimental design is to identify the significant factors with a minimum number of experiments. These identified factors are then optimized to particular levels according to the design. Subsequently, they are confirmed by performing the optimized experiments. This paper is to outline the experimental design for these complex and time consuming experiments. It will be extremely useful in determining control factors, optimizing formulations, and screening chemicals.

EXPERIMENTAL

DEFINING FACTORS AND LEVELS

The quality of the carbon black pigment dispersion is controlled by many factors. The grade of carbon black used will affect the results significantly. Black Pearls® 490, a medium grade carbon black from Cabot Corporation (Billerica, MA) was used. It is designed for general printing ink applications and some of its properties are listed in Table 1. It has relatively small particle size and medium surface area which contributes to its dark mass-tone and strong tinting strength. The smaller and more uniform the particle size, the greater the jettness and tinting strength. Conversely, the larger the particles, the grayer is the mass-tone of the black. Another key property of carbon black, besides particle size, is the morphology of the primary aggregate, known as "structure".

Black Pearls® 490 is a high structure black with its primary aggregates composed of many prime particles containing branches and chains. This affects the packing of the pigment and make it easier to disperse. Furthermore, low structure black particles are too small to interfere with the longer wavelengths of the visible light spectrum. When mixed with white pigment, the resultant gray has a marked red-brown tone. On the other hand, high structure carbon blacks with fewer particles per unit volume or weight, do not interfere with the shorter wavelengths of light. This results in a clean blue undertone color. The effects of particle size and structure of

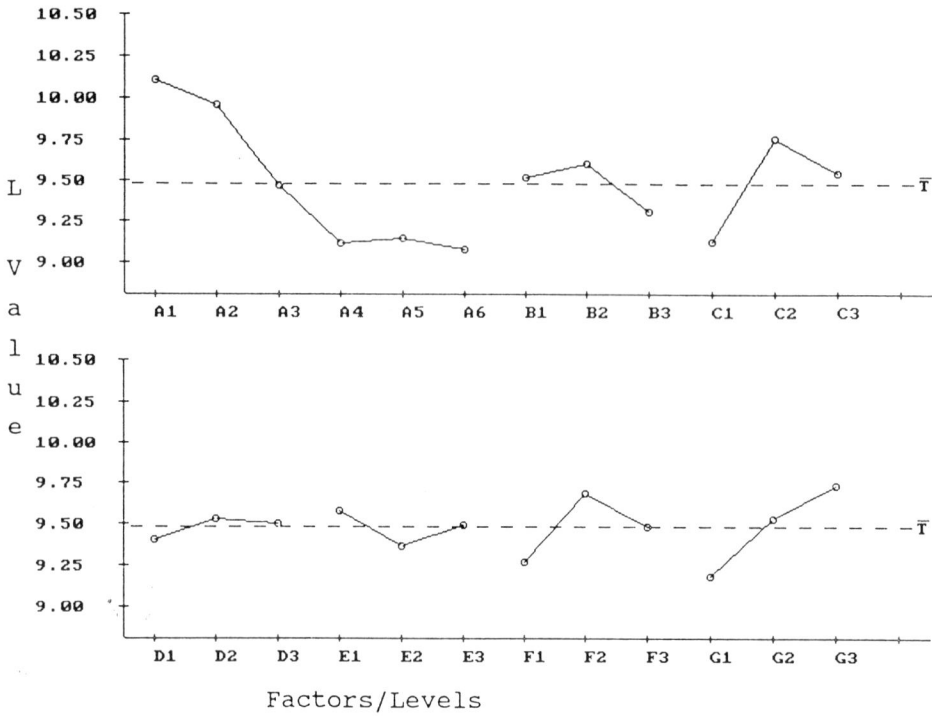

Figure 1. Linear Graphs of L-Value Versus Factors/Levels.

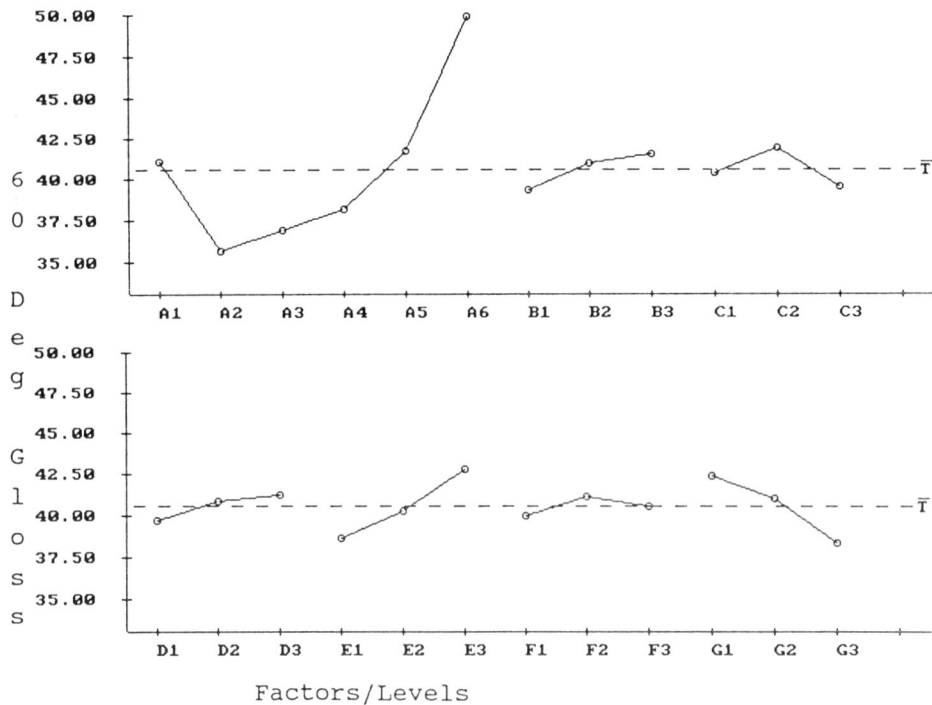

Figure 2. Linear Graphs of 60 Degree Gloss Versus Factors/Levels.

carbon black pigments are summarized in Table II. Since the objective of this paper is to study the effectiveness of polymers in pigment dispersion, this factor or the choice of different grades of pigment was kept constant.

The first controllable factor is the type of styrene/acrylic copolymer used in the pigment dispersion. Three resins, Polymer A, B, and C were used. Their properties are listed in Table III. They have high acid number and therefore, are alkaline soluble. The main difference between them is the molecular weight. All of them are in solid form. When dissolved in basic solution, the viscosities vary accordingly. These three copolymers themselves are the three levels for this factor.

Factor two is the type of acrylic polymers used. The three levels are Polymer D, E, and a 50/50 mixture of the polymers. The typical properties of the Polymer D and E are also listed in Table III. These polymers have lower acid number but higher molecular weight than the styrene/acrylic copolymers. Polymer D and E also differ in molecular weight. This type of polymer has dual functions as both pigment grinding resin and as the letdown vehicle in low cost printing inks.

The mix ratio of the styrene/acrylic copolymers and the acrylic polymers is the third factor. Instead of studying an individual type of polymer, this will determine the compatibility of the polymers and enable us to optimize the blend ratio for best performance. Six levels at: 0.0/1.0, 0.2/0.8, 0.4/0.6, 0.6/0.4, 0.8/0.2 and 1.0/0.0 were the mix ratios of the copolymers and acrylic polymers.

Table I. Physical properties of Black Pearl® 490

PHYSICAL PROPERTIES	ESTIMATED VALUE
Surface Area	87.0 m^2/g
Particle Size	25.0 nM
Oil Absorption	124.0 cc/g
Density	21.0 lb/ft^3

Table II. Typical Properties Related to Carbon Black Partical Size and Structure (Reprint from Cabot's Technical Report S-136 with Permission)

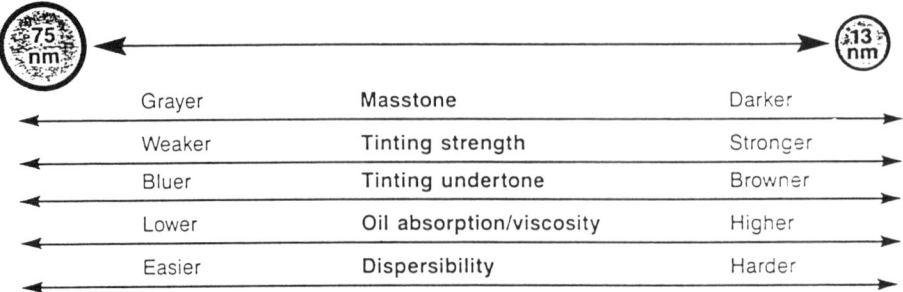

Grayer	Masstone	Darker
Weaker	Tinting strength	Stronger
Bluer	Tinting undertone	Browner
Lower	Oil absorption/viscosity	Higher
Easier	Dispersibility	Harder

Typical properties as related to carbon black structure (aggregate size).

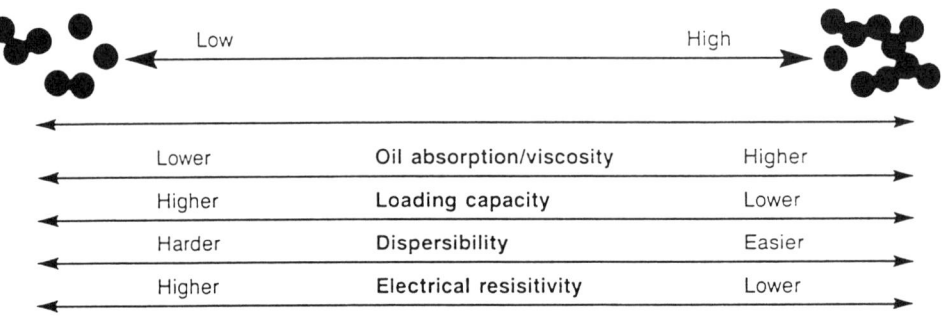

Lower	Oil absorption/viscosity	Higher
Higher	Loading capacity	Lower
Harder	Dispersibility	Easier
Higher	Electrical resisitivity	Lower

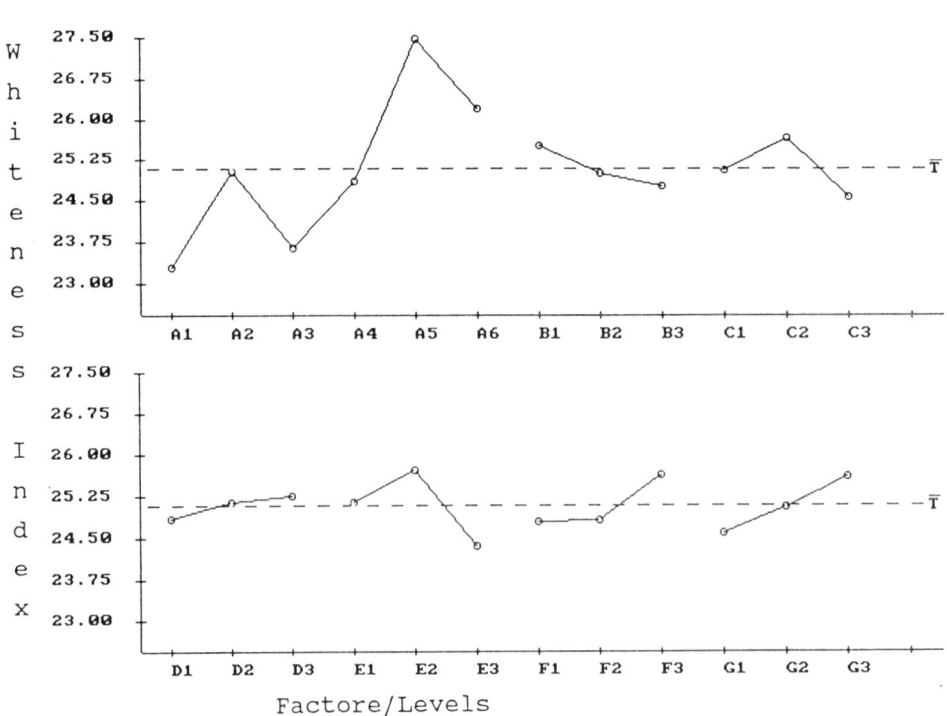

Figure 3. Linear Graphs of Whiteness-Index Versus Factors/Levels.

Factor four is the pigment to binder ratio. This is an important factor as it will show the efficiency of the resins. Since pigment particle size is an important geometric factor, the pigment volume concentration (PVC) is used more often than the weight concentration, especially when comparing different kinds of pigments. When a binder is added to a pigment, the first stage is the binder progressively displacing air from within the interstices of the packed bed of pigment. During this stage, pigment, binder and air exist in the system. Finally, a point is reached, where the last trace of air is displaced and the voids of the pigment are completely filled with the binder. This unique point is called the critical pigment volume concentration[4] (CPVC) and can be determined by different experimental methods. The most commonly used method is to obtain the oil absorption value defined by ASTM D281-31 and D1483-60). Further addition of binder leads to a second stage where the pigment particles start to separate from each other. Below the CPVC, pigment particles lose contact, whereas, above the CPVC, binder is replaced by air. Generally, a pigment to binder ratio of four is being used in commercial pigment dispersions for printing inks, well below the CPVC to assure stability. Three levels at 5/1, 4/1, and 3/1 were used in this study.

The fifth factor is the grinding time. A specific grinding time is required for different equipment. Since most equipments require high energy to disperse pigments, it will be desirable to have a shorter grinding time with equivalent quality of dispersions. Too short a grinding time will result in big aggregates of pigment, low gloss, unstable dispersion, and low color strength. On the other hand, over grinding the pigment will break the pigment structures, creating new surface area. Demand of dispersing resin may increase and lead to instability. Damage to the grinding shot or the milling equipment can occur also. Using our laboratory shot mill, the three levels of grinding were 10, 20 and 30 minutes respectively.

Defoamer is the sixth factor. Foaming is always a problem in manufacturing pigment dispersions and inks. It can also lead to a problem on the printing press. Insufficient defoamer may cause a foaming problem which can result in a low quality pigment dispersion. Excess defoamer will make wetting the pigment difficult and affect the gloss or the effectiveness of the dispersing resin. A hydrophobic silica defoamer was used and the three levels were 0.0%, 0.25% and 0.5%.

The last factor is the surfactant. Surfactants or wetting agents are widely used in printing inks, especially in water based systems. The surface tension of water is high (72 dynes/cm) comparing to organic solvents in the range of 20 to 30 dynes/cm (i.e. toluene = 28 dynes/cm, n-propanol = 24 dynes/cm). Incorporating surfactant(s) in a water based system will lower the surface tension and enhance the wettability of the pigment. Nevertheless, excess surfactant will almost certainly cause a foaming problem. An alkyl sulfosuccinate surfactant at 0.0%, 0.5%, and 1.0% levels were used.

Table III. Physical Properties of the Grinding Resins

POLYMER PROPERTIES	STYRENE/ACRYLIC COPOLYMERS			ACRYLIC POLYMERS	
	A	B	C	D	E
MOLECULAR WEIGHT (M_n)	1800	2600	4500	25000	42000
ACID NUMBER	210	215	155	110	135
T_g (°C)	65	100	60	30	40
PHYSICAL FORM	ALKALINE SOLUBLE SOLID			TRANSLUCENT SOLUTION	

The above factors and their respective levels are summarized in Table IV. If a "one variable/one level" approach was used to examine all the possible mentioned factors and levels, the total number of experiments would be 6×3^6 or 4374.

TAGUCHI'S EXPERIMENTAL DESIGN

Taguchi[1,2] developed a family of fractional factorial experiment (FFE) matrices which can be utilized in various situations. FFEs are designed statistically such that only a portion of the total possible combinations are used to estimate the main factor effects and some interactions. The invention of orthogonal arrays was recorded as early as 1897 by Jacques Hadamard, a French mathematician. The Hadamard and Taguchi matrices are identical mathematically with the columns and rows rearranged. Analysis of variance (ANOVA) was developed by Sir Ronald Fisher in the 1930s. It is the statistical method used to interpret experimental data and detect any differences in average performance of groups of items tested. The selection of an orthogonal array depends on the number of factors and levels selected. Two-level arrays are L4, L8, L12, L16, L32, etc. Three-levels arrays are L9, L18, L27, etc. The number in the array designates the number of experiments. From ANOVA, the minimum required degrees of freedom in the experiment is the sum of all the factors and the interaction degree of freedom.

For this study, there are seven factors with one factor at six levels and six factors at three levels as mentioned previously. The L18 ($2^1 \times 3^7$) orthogonal array was selected as shown in Table V. Regular L18 can only handle one factor at two levels and seven factors at three levels each. It was modified by combining two columns into one so that one factor can contain six levels (Table VI). The actual 18 experiments performed are outlined in Table VII. The total degrees of freedom available in an orthogonal array is equal to the number of experiments minus one. Details of the experimental design can be found in the literature[1,2,3].

GRINDING PROCEDURE

Different methods and types of equipment used for pigment dispersions in graphic arts are well known in the industry[5,6]. They will not be discussed in detail in this paper. Since the pigment/binder ratios change in these experiments, the grindings were done at constant solids (37.5%) rather than constant pigment loadings. This will keep the dispersion viscosities relatively close to each other and eliminate this variable. When evaluating the quality of the pigment dispersions, they were readjusted to the same pigment levels for equal comparison. A total of 600 grams for each experiment was weighed carefully. A high speed laboratory disk disperser

Table IV. Selected Factors and Their Levels

SERIAL NUMBERS	FACTORS	NUMBER OF LEVELS	LEVELS
A	Copolymer/Polymer Ratio	6	0/1,2/8,4/6,8/2,1/0
B	Styrene/Acrylic Copolymer Type	3	Polymer A, B, C
C	Acrylic Polymer Type	3	Polymer D, E and 50/50
D	Pigment/Binder Ratio	3	5/4, 4/1, 3/1
E	Grinding Time (Minutes)	3	10, 20, 30
F	Defoamer (%)	3	0.0, 0.25, 0.5
G	Surfactant (%)	3	0.0, 0.5, 1.0

Table V. Taguchi Orthogonal Array - L18 (2^1 x 3^7)

EXPERIMENT NUMBER	FACTORS							
	1	2	3	4	5	6	7	8
1	1	1	1	1	1	1	1	1
2	1	1	2	2	2	2	2	2
3	1	1	3	3	3	3	3	3
4	1	2	1	1	2	2	3	3
5	1	2	2	2	3	3	1	1
6	1	2	3	3	1	1	2	2
7	1	3	1	2	1	3	2	3
8	1	3	2	3	2	1	3	1
9	1	3	3	1	3	2	1	2
10	2	1	1	3	3	2	2	1
11	2	1	2	1	1	3	3	2
12	2	1	3	2	2	1	1	3
13	2	2	1	2	3	1	3	2
14	2	2	2	3	1	2	1	3
15	2	2	3	1	2	3	2	1
16	2	3	1	3	2	3	1	2
17	2	3	2	1	3	1	2	3
18	2	3	3	2	1	2	3	1

Table VI. Modified Taguchi Orthogonal Array - L18 (6^1 x 3^6)

EXPERIMENT NUMBER	FACTORS							
	A		B	C	D	E	F	G
1	1	0	1	1	1	1	1	1
2	1	0	2	2	2	2	2	2
3	1	0	3	3	3	3	3	3
4	2	0	1	1	2	2	3	3
5	2	0	2	2	3	3	1	1
6	2	0	3	3	1	1	2	2
7	3	0	1	2	1	3	2	3
8	3	0	2	3	2	1	3	1
9	3	0	3	1	3	2	1	2
10	4	0	1	3	3	2	2	1
11	4	0	2	1	1	3	3	2
12	4	0	3	2	2	1	1	3
13	5	0	1	2	3	1	3	2
14	5	0	2	3	1	2	1	3
15	5	0	3	1	2	3	2	1
16	6	0	1	3	2	3	1	2
17	6	0	2	1	3	1	2	3
18	6	0	3	2	1	2	3	1

(Series 2000 from Premier Mill Corp., New York, NY) was used for the initial wetting and dispersing of the pigment. The disperser was run for 30 minutes at 4000 rpm. Then, it was transferred to a laboratory scale horizontal shot mill (3 HP, Model "Mini 250" from Eiger Machinery, Inc., Mundelein, IL) to complete the grinding of the pigment dispersion. The horizontal position enables the mill to be restarted regardless of the product viscosity or the type of grinding media. The type of grinding media is not a limiting factor anymore. As in the case of steel shots, it may cause a problem in vertical mills (settlement on the bottom). The laboratory shot mill's empty chamber volume is 250 ml. 1 mm glass shot were used as the grinding media to fill the chamber. Void volume was estimated to be 100 ml. The mill was run at 3500 rpm and drew 0.5 amp.

PIGMENT DISPERSION EVALUATION PROCEDURE

Fineness of grind: The degree to which the black pigment was dispersed was checked by using a grind gauge[7,8]. ASTM test method D1210-64 defines a conventional grind gauge procedure to be used. The Hegman scale, sometimes referred to as the National or North Standard scale, was used. This method gives no information regarding the actual degree of dispersion. It gives only a useful idea of the number and size of the large or undispersed particles. It is incapable of providing data on the overall reduction in particle size distribution that is occurring. Usually, as the time of grinding increases, the number of clusters decreases. It is assumed that the general degree of the dispersion of the finer particles has also improved. This test provides a quick, inexpensive, and reproducible procedure for the pigment dispersion. This method becomes an indispensable tool and is widely used in graphic arts industry.

Re-adjustment: The pigment/binder ratios and pigment loadings in the dispersions were different while the solids level was kept constant. In order to compare the dispersions equally, all of the dispersions were adjusted to 20% pigment at the pigment/binder ratio of 3/1 by post-adding polymers or water. A #5 Meyer's rod was used to draw down the dispersions for equivalent coating weight. The samples were then dried in an oven at 80 °C for 15 seconds.

Jettness/L-value: The adjusted black pigment dispersions were coated on uncoated paper. A color meter (Base Unit Model #4411, Color Meter head Model #4431 from BYK-Chemie, Wallingford, Conn.) was used to read the L-value. All colors can be defined in the three dimensions of color space. The most commonly used color space is known as Commission Internationale de L'Eclairage L-a-b system (CIELab). It is defined by a vertical axis which is the lightness/darkness (white/black) axis, designated as the L-value. The more

Table VII. Actual Factors and Levels of the Modified Taguchi Orthogonal Array - L18 (6^1 x 3^6)

EXP. NO.	FACTORS						
	POLYMER RATIO	COPOLYMER TYPE	POLYMER TYPE	P/B RATIO	GRIND TIME (MIN)	DEFOAMER (%)	SURFACE ACTIVE AGENT (%)
1	0/1	A	D	5/1	10.0	0.00	0.0
2	0/1	B	E	4/1	20.0	0.25	0.5
3	0/1	C	50/50	3/1	30.0	0.50	1.0
4	2/8	A	D	4/1	20.0	0.50	1.0
5	2/8	B	E	3/1	30.0	0.00	0.0
6	2/8	C	50/50	5/1	10.0	0.25	0.5
7	4/6	A	E	5/1	30.0	0.25	1.0
8	4/6	B	50/50	4/1	10.0	0.50	0.0
9	4/6	C	D	3/1	20.0	0.00	0.5
10	6/4	A	50/50	3/1	20.0	0.25	0.0
11	6/4	B	D	5/1	30.0	0.50	0.5
12	6/4	C	E	4/1	10.0	0.00	1.0
13	8/2	A	E	3/1	10.0	0.50	0.5
14	8/2	B	50/50	5/1	20.0	0.00	1.0
15	8/2	C	D	4/1	30.0	0.25	0.0
16	1/0	A	50/50	4/1	30.0	0.00	0.5
17	1/0	B	D	3/1	10.0	0.25	1.0
18	1/0	C	E	5/1	20.0	0.50	0.0

Table VIII. Optimized Formulae from the Taguchi Design

FORMULA # 1 - FOR HIGH JETTNESS AND GLOSS						
FACTORS	A	B	D	E	F	G
	POLYMER RATIO	STYRENE/ ACRYLIC COPOLYMER	P/B	GRIND TIME	DEFOAMER	SURFACTANT
LEVEL	6	3	2	3	1	1
	0/1	POLYMER C	4/1	30.0 MIN.	0.0%	0.0%

FORMULA # 2 - FOR HIGH COLOR DEVELOPMENT						
FACTOR	A	C	D	E	F	G
	POLYMER RATIO	ACRYLIC POLYMER	P/B	GRIND TIME	DEFOAMER	SURFACTANT
LEVEL	1	1	2	3	1	1
	1/0	POLYMER D	4/1	30.0 MIN.	0.0%	0.0%

positive the L-value, the whiter is the color. As in the case of the black pigment dispersion, the smaller the L-value, the darker or the jetter is the black. The other two horizontal axes are green/red (a-value) and yellow/blue (b-value). These two values are not important in this study as the color of choice is black.

Gloss: The 60 degree gloss from the same bar-down samples as mentioned above was measured by a gloss meter (Base unit Model #4411, Gloss Meter head Model #4344, BYK-Chemie, Wallingford, Conn.). Gloss is an appearance quality of the specularly (mirror-like) reflected light. Optically, gloss is a geometric attribute of the reflected light by surfaces. Every surface reflects some diffused light in all directions and some specular light at an angle equal to the angle of incident light. Black surfaces reflect very little specular light. Therefore, any incompatability or defects in the black pigment dispersions will be picked up by the gloss measurement. The conventional glossmeter[9] measures the intensity of specularly reflected light disregarding the distinctness of the reflected image. By contrast, the human eye is very sensitive to distinctness of image, rather than to its intensity[10]. The intensity of specularly reflected light depends on the angle of observation and on two characteristics of the reflecting surface, its refractive index and its roughness. Both pigment and the grinding resin contribute to surface roughness and refractive index, hence, the gloss.

Color Strength/Whiteness-Index: In order to determine the color development, all adjusted pigment dispersions were mixed with a predispersed white tint, Ti-Pure R940 (from DuPont, Wilmington, DE) at the ratio of 5 to 95 (black to white). This is known as bleaching and is commonly used in printing ink industry. They were then coated on Leneta 3NT-3 paper. The whiteness-index was recorded using the same color meter as before. The whiteness-index is similar to the L-value, the smaller the number, the darker the black dispersion. The whiteness-index is usually not sensitive enough to measure full strength dispersions or inks. Bleaching provides a method to dilute the color consistently and enables one to measure the color strength. Sometimes, one can find the difference in color strength by visual observations using this method.

All the measured data for the 18 experiments were collected and fed into a computer for analysis.

RESULTS AND DISCUSSION

For the fineness of grind, all the black pigment dispersions had a value of 7 or higher on the Hegman scale. This translates to particle size of 0.5 mil or smaller. It indicates that there are no large or undispersed pigment agglomerates and the dispersion is completed. Since no difference is found for the fineness of grind, it is not included in the computer analysis.

Table IX. Comparison of the Results from the Optimized Formulae of the Taguchi Predictions and the Confirmation Experiments

THEORY/ EXP. DATA	L-VALUE		GLOSS		WHITENESS-INDEX	
	FORMULA #1	FORMULA #2	FORMULA #1	FORMULA #2	FORMULA #1	FORMULA #2
TAGUCHI	8.48	9.32	54.64	44.6	24.49	21.88
PREDICTION	+/-0.52		+/-5.64			
EXPERIMENT	8.37	9.22	54.0	39.6	25.07	22.19

The Taguchi methods/ANOVA computer program (ANOVA-TM, Version 2.2, Advanced Systems & Designs, Inc., Farminging Hills, MI) was used. The software was installed and run from a personal computer. This will enable one to use the Taguchi's design and analyze the data without the need to know the mathematics and statistics in detail. The measured L-value, 60 degree gloss, and whiteness-index were all fed into the computer. The actual data for the 18 experiments will not be shown here because they are difficult to analyze by just looking at them. Using the statistical analysis technique ANOVA computer program, linear graphs for each factor at different levels were plotted against the property measured by the computer. The curves were shown in Figures 1, 2 and 3 for L-value, 60 degree gloss and white-index respectively.

Figure 1 shows the linear graphs of the L-value versus the seven factors at their designed levels. Remember that the lower the L-value, the jetter is the black dispersion. The L-value decreases with increase in styrene/acrylic copolymer to acrylic polymer ratio (factor A). It reaches a minimum at 4/6 ratio and then levels off. This shows that the more the styrene/acrylic copolymer used, the jetter is the black dispersion. The strong interaction suggests that the two types of polymers may not be compatible. Out of the three styrene/acrylic copolymers (factor B), Polymer C, the highest molecular weight one, gives the lowest L-value, i.e., highest jettness. Polymer B and C are almost equal in jettness. For the acrylic polymers (factor C), Polymer D produces higher jettness than Polymer E with the 50/50 mixture (level 3) in between the two. The pigment to binder ratio (factor D) does not have any major effect in this study. For the grinding time (factor E), 20 minutes seems to be somewhat better than 10 or 30 minutes. The defoamer (factor F) has a negative effect on the jettness. The surfactant (factor G) used also lowers the jettness of the grind.

Linear graphs for the 60° gloss were plotted in Figure 2. For the factor A, the polymer ratio, higher gloss is obtained at the extremes, with lowest gloss at a ratio of 2/8. Again, a ratio of 1/0 (all styrene/acrylic copolymer) produces the highest gloss. Polymer C (level 3) gives the highest gloss of the three copolymers. Polymer E is just very little higher in gloss than Polymer D and the mixture of the two. The pigment to binder ratio (factor D) and the defoamer (factor F) have no effect on the gloss. Grinding the dispersion for 30 minutes results in the highest gloss. More surfactant used will further reduce the gloss.

The whiteness-index linear graphs were shown in Figure 3. Similar to the L-value, the lower the white-index, the higher is the color strength. There is some noise picked up for factor A, the polymer ratio. Nevertheless, the trend is that the lower the ratio (the less of the styrene/acrylic copolymer), the better the color development is going to be. Polymer C is slightly better than the other two copolymers while Polymer D produces higher strength than Polymer E. Pigment to binder ratio is not a factor again. 30 minutes grind results in highest strength. Addition of either the defoamer or the surfactant yields lower color strength.

From the three properties measured, one can observe the trend of the results. The styrene/acrylic copolymers and the acrylic polymers are not really compatible. When mixed together, a minimum in properties is always found in jettness, gloss and color strength. Results suggest that for high jettness and gloss, one should use styrene/acrylic copolymer with Polymer C being the best one. On the other hand, if you are looking for the highest color development possible, an acrylic (e.g 100% acrylic without styrene) polymer is the choice and Polymer D produces better color than Polymer E.

The quality of the pigment dispersion is not a function of the pigment to binder (P/B) ratio. The chosen range was from 5/1 to 3/1 only. Perhaps the range is too narrow to observe any effect. However, it is uncommon to grind pigment outside this range. A higher ratio will approach the critical pigment volume concentration and poorer dispersion will be expected. A lower ratio is not practical as pigment processed per batch will be too low and it will not be economical. Results actually indicate that the polymers are effective grinding resins and have a wide useable range.

Grinding time (from 10 to 30 minutes) has a relatively very small effect on the pigment dispersions. The Taguchi result agrees with the data from the fineness of grind. All dispersions reached a very fine dispersion already. Extra milling will not guarantee a better dispersion and in certain cases, overgrinding will actually make the dispersion worse. From the laboratory mill configuration, it was calculated that the average residence times were 2.5, 5.0 and 7.5 minutes for grinding times of 10, 20 and 30 minutes respectively. The resins were able to complete the dispersions effectively in a relatively short residence time.

In this study, it was found that addition of defoamer and surfactant did not improve the dispersions. They actually had an adverse effect on the dispersions. Defoamer works when it is almost incompatible with the system so that it will stay at the liquid/air interface. The surfactant used is probably not compatible with the defoamer. Compatibility is very critical with black dispersions. Slight incompatability will result in lower jettness, gloss and color strength.

OPTIMIZATION AND CONFIRMATION

Optimizing From The Design: From the results of the Taguchi's experimental design and the linear graphs of the ANOVA, optimized formulae and process can be obtained. They are summarized in Table VIII. This was done by simply picking the best level for each factor from the linear graphs. In the case that the particular factor had no significant effect, the mid-level was used.

In this study, two different main properties were optimized resulting in two formulae. The first formula is optimized for high jettness (low L-value)/60 degree gloss. The factors and the levels for the first formula are as follows: A6/B3/D2/E3/F1/G1. This means that highest jettness and gloss can be obtained by using an all styrene/acrylic copolymer (A6); using Polymer C (B3); a pigment:binder at 4:1; grinding for 30 minutes (E3); and using no defoamer nor surfactant (F1, G1). Since an all styrene/acrylic copolymer is used, the choice of factor B (acrylic polymer) and its level (choice of polymer) has no meaning.

The second optimization was done for high color development (low whiteness-index). The formula is: A1/C1/D2/E3/F1/G1. For highest color strength, one should use an all acrylic polymer (A1); Polymer D (C1) is the choice; a pigment/binder ratio at 4/1; grinding for 30 minutes (E3); using no defoamer (F1) and no surfactant (G1). As an all acrylic polymer system is recommended, factor B (styrene/ acrylic copolymer type) and its level is not included in the formula.

Confirmation Experiments And Their Results: The optimized formulae are predicted from the experimental design using statistical methods. However, it does not guarantee that the predictions will work. The only way to confirm this is to perform the confirmation experiments on the optimized formulae. The ANOVA-TM computer program used is able to predict the values of L-value, 60 degree gloss and whiteness-index by inputting any levels for each factor based on the 18 experiments. The results of the prediction and the actual experimental data for the optimized formulae are summarized in Table IX. Formula #1 was optimized for jettness/gloss and formula #2 for color development. In general, the design predictions agree with the confirmation experimental data. The design predicts L-value and gloss for formula #1 as 8.48 and 54.64 respectively. The actual data obtained was 8.37 and 54.0 which are very close to the prediction. For formula #2, predicted value for whiteness-index is 21.88 with confirmation experimental data as 22.19. Again, the results agree. This confirms that the formulae are optimized. By performing 20 experiments (18 from the design and 2 for the confirmation), the Taguchi experimental design, with the help of the computer software, was able to provide enough information to determine: 1) which are the significant factors, 2) and at what levels should they be used. This eliminates the need to perform over four thousand experiments for all possible combinations while still being able to obtain the optimum formulations and the influence of the factors.

CONCLUSIONS

Various direct and indirect compositional factors on pigment dispersion were studied efficiently utilizing the Taguchi Orthogonal Array Technique. The two type of polymers have distinct properties. Styrene/acrylic copolymers produce high jettness and gloss for the black pigment dispersions. These are important properties for surface printing as in packaging inks for flexible substrates. On the other hand, acrylic polymers develop high color strength for the black pigment, a more important factor than jettness and gloss for corrugated and news inks. Moreover, the acrylic polymers used in this study can be used as the letdown vehicle also. They can be formulated as a single polymer ink system. This makes raw material handling, storage and QC easier for ink manufacturers. Therefore, depending on the end use, specific polymers can be recommended for different applications.

The five water based polymers all produced good black dispersions effectively within the pigment to binder ratio (3/1 to 5/1) and the grinding time (10 to 30 minutes) constraints. Polymer C produces slightly better jettness and gloss than polymer A and B, but the other two polymers being at lower molecular weight (i.e. lower solution viscosity), can be used in formulations that required higher solids. While polymer D develops higher color strength than polymer E, polymer E can reduce the raw material cost of the finished ink at same viscosity but lower solids.

The defoamer and surfactant used in this study were not compatible with the system. In search for the right products, it may be time for another Taguchi experimental design to screen chemicals. The Taguchi OA strategy was shown to be effective in reducing experimental effort (i.e. time and cost) without losing any valuable information. This technique shall be used for wider applications to optimize products and process variables as well as to increase productivity in the printing ink industry.

ACKNOWLEDGMENT

The author would like to thank P. N. Murphy for his thoughtful discussions and R. A. Snow and P. C. Marcella for performing the experiments. The author is also indebted to W. R. Furlan for his useful discussions on the Taguchi design and assistance in the analysis of the data.

REFERENCES

1. Taguchi, G., and Konishi, S., Taguchi Methods, Orthogonal Arrays and Linear Graphs, Am. Supplier Institute, Inc., Michigan, (1987).

2. Taguchi, G., System of Experimental Designs, Am. Supplier Insitute, Inc., Michigan, (1988).

3. Ross,P.J., Taguchi Techniques for Quality Engineering, McGraw Hill Publishing Co., NY,(1988).

4. Asbeck, W.K., and Maurice, V.L., Critical Pigment Volume Relationships, Ind. Eng. Chem., **41(7)**, 1470-1475, (1949).

5. Parfitt, G.D., Dispersions of Powders in Liquids; with Special Reference to Pigments, 3rd Ed., Applied Science Publisher, NJ, (1981).

6. Patton, T.C., Paint Flow and Pigment Dispersion, 2nd Ed., John Wiley & Sons, (1987).

7. Patton, T.C., Sieving, Pigment Handbook, Vol. III, Wiley-Interscience, NY, (1973).

8. ASTM Standard Test Method D1210-64, Test for Fineness of Dispersion of Pigment-Vehicle System, (1970).

9. Hunter, R.S., The Measurement of Appearance, John Wiley & Sons, NY, (1975).

10. Harrison,V.G.W., Definition and Measurement of Gloss, The Printing and Allied Trades Research Association, Cambridge, England, (1945).

ADVANCES IN WATERBORNE ACRYLIC RESIN/POLYESTER

POLYMER BLENDS AND THEIR APPLICATIONS: PART-A

Mahendra K. Sharma

Research Laboratories
Eastman Chemical Company
P. O. Box 1972
Kingsport, TN 37662 (USA)

ABSTRACT

Recent developments in waterborne polymers and polymer blends are discussed in relation to their use as a binder in several applications such as coatings, paints and ink systems. A process for the preparation of waterborne acrylic resins/polyester polymer blends that avoids the need for surfactants is described. It was found that these acrylic and polyester polymer blends can not be prepared without using this process. The stable polymer blends can be prepared with acrylic resins/polyester polymer ratios in the range of 10/90 - 35/65 (wt/wt). The higher solid content can be achieved by either incorporating low molecular weight acrylic resins or by adding 5.0 -10.0 wt% alcohol. These polymer blends exhibit low viscosity (e.g. 40-200 cps) with 30 wt% solid content. The viscosity of the polymer blends can be adjusted by suitable thickener for a given application. Results indicated that the acrylic/polyester polymer blends exhibit superior pigment grinding and film properties as compared to the polyester or acrylic polymer alone.

INTRODUCTION

The demand for waterborne polymers and/or polymer blends is growing due to new environmental and health regulations at local, state and federal levels, which restrict the use of solvents in paints, coatings and printing processes. These regulations and consumer demands meet or exceed an existing

Surface Phenomena and Latexes in Waterborne Coatings and Printing Technologies, Edited by M.K. Sharma, Plenum Press, New York, 1995

41

quality obtained from solvent-borne coating systems, and are forcing the industries to develop environmentally friendly products.[1-5] Among several options to develop technology, water is the best choice to use as a carrier in formulating coating systems. In order to reduce or eliminate organic solvents from formulations, the solvents should be replaced partially or completely with an environmentally safe solvent (e.g. water), in the coating formulations to achieve either a zero or low volatile organic compounds (VOCs) content. The possible methods of eliminating or minimizing organic solvents in coating and paint formulations include the following:

1. The coatings, paints and printing systems employ a polymer solution as a binder in a mixed organic and aqueous solvent. This method only partially eliminates the need for organic solvents.

2. The coating systems employ an aqueous polymer solution as a binder. This method is limited to water soluble polymers and fails to provide water resistance to the finished products. In addition, a large amount of water needs to be removed, which limits its use in some applications such as textile sizing and printing, where the speed of coating and printing processes must be reduced due to slow drying of the coating material on substrates.

3. These coating systems involve an aqueous solution of alkali salts (e.g. sodium, potassium, ammonium etc.) of polymers used as a binder. Among various types of polymers, acrylic based polymers are widely used in formulating waterborne products. If ammonium salt of acrylic polymer is used, the change in pH of the product due to escape of ammonia creates several problems in the process as well as affects end-use properties of the films. If other salts like sodium, potassium etc. are used, the finish products are water sensitive due to the presence of these salts in the polymer.

4. These coatings, paints and printing systems involve a blend of the polymers as a binder. In general, the properties of coating/printing ink films are significantly improved as compared to a system without the polymer blends.

Several water-soluble and water-dispersible polymers are available commercially. The water soluble polymers have very limited applications. These polymers are widely used as additives like rheology modifiers. Because of water solubility, the dried coating film can not provide water resistance. The water dispersible polymer can provide a water resistant film, and can be used in a wide variety of coating and printing applications. The fast drying can be provided by dispersible polymers during coating and printing processes as compared to water-soluble polymers due to the dispersible nature of the polymers. The water can easily be removed from the dispersions as it is physically adsorbed on the polymer surface. In case of polymer solutions, the water interacts with polymer at the molecular level resulting in a slow drying of the film. The polymers used in organic coatings and printing formulations are described previously.[6-13]

WATER-SOLUBLE POLYMERS

The polymers which produce a clear solution in water and do not scatter light are called water-soluble polymers. The polymer molecules in the solvent medium are randomly distributed. The random distribution of these polymer molecules depends on the concentration, molecular size and shape, as well as their interaction with solvents.

Both naturally and synthetically available polymers are used in coating and printing processes. The naturally available polymers used in coating processes are alignates, dextrins and starches, whereas the synthetic polymers include polyvinyl alcohol (PVA), polyvinyl methyl ethers (PVME), polyacrylamides (PAA), salts of polyacrylic acids (SPAA), and ethylene oxide based polymers.

A large number of polymers can be made water soluble either at low or at high pH. The polymers soluble at a pH less than 7 (acid medium), include copolymers composed of diethylaminoethyl acrylate, 2-vinyl pyridine, t-butyl aminoethyl methacrylate, and dimethylaminoethyl methacrylate. The polymers soluble at a pH above than 7 (alkaline medium), include acrylic copolymers, alpha proteins, acidic polyesteramides, styrenated shellac, styrene-maleic copolymers, rosin-maleic and their ester based polymers.

WATER-DISPERSIBLE POLYMERS

The water-dispersible polymers are most promising for formulating waterborne coatings and printing inks. Since these polymers are in dispersion form, low viscosity can be achieved at a high solids content as compared to water soluble polymers. Hence, one can formulate a desired coating/printing ink system with a high solids content in the presence of water as the solvent.

Among several methods for preparing water-dispersible polymers such as controlled flocculation, solvent extraction, surface modification and emulsion polymerization, polymerization of monomers in the presence of emulsion is widely used to form latexes. The latex particle size is usually in the range of 50 - 400 nm. These latex systems are fairly stable, and can form a film with and without a coalescing agent depending on the properties of the dispersed polymer. The polymers prepared by emulsion polymerization include acrylic polymers and copolymers, vinyl polymers and styrene-acrylic copolymers. Based on monomer composition and type, one can prepare polymers with desired end-use properties of the coating/printing films.

The polymers with high solid content at low viscosity provide a high coating weight deposited from a relatively thin wet film. The high molecular weight latexes generally form harder, excellent heat resistant, abrasion resistant, water resistant films and provide better adhesion than that of the low molecular weight polymers.

The water-soluble and water-dispersible polymers provide both advantages and disadvantages during the film formation. In general, water soluble polymers exhibit poor water resistance, while water-dispersible polymers exhibit poor gloss, film formation and adhesion. Therefore, it is planned to form polymer blends containing water-soluble acrylic and water-dispersible polyester polymers, in order to achieve improved film properties, which can not be obtained either by water-soluble or water-dispersible polymers alone.

EXPERIMENTAL

MATERIAL

The acrylic resins and polyester polymers used to form waterborne polymer blends are described in Table I. Normal propanol or isopropanol was used in certain polymer blends to obtain the desired viscosity of the waterborne polymers. Deionized water was used through out the experiments.

METHOD

Acrylic resins were dissolve in an aqueous alkaline medium during continuous stirring of the resin flakes. Ammonium hydroxide was used to prepare the alkaline solution. The alkaline resin solution was heated to about 80°C in order to remove excess ammonia (e.g. pH less than 8.5). At 80°C, the polyester pellets were added slowly to the alkaline resin solution during stirring. The temperature was kept at 80°C until the polyester pellets were dispersed in the solution. The polyester blends were cooled to about 50°C. If the viscosity of the polymer blend is high, about 5-10 wt% n-propanol or isopropanol was added to the mixture. The material was cooled to room temperature with continuous stirring, and was stored in the plastic/glass container. The process of preparing these polymer blends[14] are schematically shown in Figure 1.

RESULTS AND DISCUSSION

The acrylic/polyester blends prepared according to the above described process are divided into two groups, based on the alcohol content. It was found that certain acrylic resin used for the preparation of the acrylic/polyester blends does not require alcohol to achieve low viscosity of the aqueous polymers. The composition and properties of these polymer blends are discussed as follows:

ACRYLIC/POLYESTER BLENDS WITH ALCOHOL

Acrylic resins belonging to this group include Joncryl-67, Joncryl-678 and Joncryl-680, as these require alcohol to make low viscosity polymer blends. The properties of these polymer blends depend on acrylic resin and polyester polymers. It was found that stable blends could be prepared by using

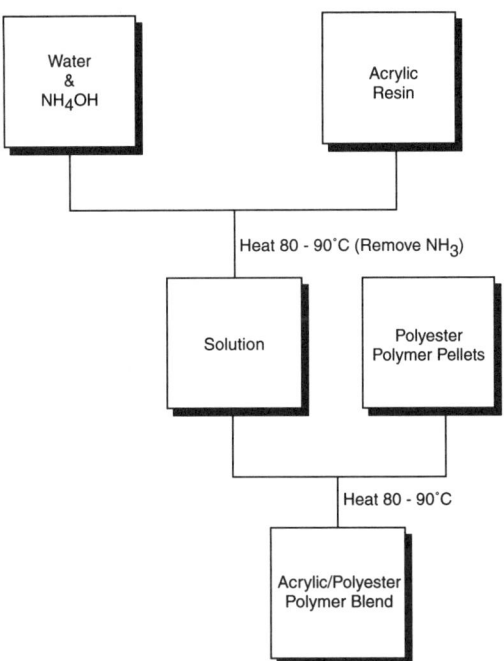

Figure 1. A Schematic Illustration of the Process
for Preparing Waterborne Acrylic/
Polyester Polymer Blends.

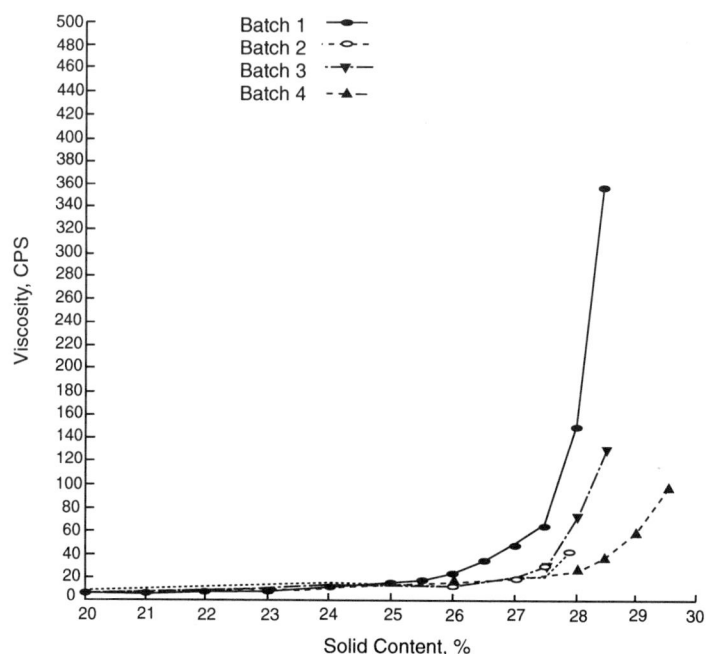

Figure 2. Viscosity as a Function of Solid Content
for Several Batches of Waterborne
Acrylic/Polyester C Polymer Blends.

TABLE I. Polymers Used for Preparing Waterborne Acrylic/Polyester Polymer Blends

ACRYLIC/POLYESTER PROPERTIES	ACRYLIC RESINS/POLYMERS					POLYESTERS		
	(*)J-67	J-678	J-682	J-680	J-683	A	B	C
Molecular Weight (M_w)	12,500	8,500	1,700	4,900	8,000	14,000	18,000	18,000
Acid Number	213	215	238	215	160	<2	<2	<2
Glass Transition Temperature (T_g, °C)	73	85	56	67	75	29	38	55
Appearance	Flakes	Flakes	Flakes	Flakes	Flakes	Pellets	Pellets	Pellets

(*)J = Joncryl (Acrylic Resins)

46

polyester C, while polyester A and polyester B did not form stable polymer blends. The composition of the waterborne acrylic/polyester blends is outlined in Table II.

The non-volatile content in the aqueous polymer blend was 28.5 weight percent. The GC analysis showed that the polymer blends contained 9.88 wt% n-propyl alcohol. The I.V. of the

Table II. Composition of Waterborne Acrylic/Polyester Blends in the presence of Alcohol

INGREDIENTS	AMOUNT	
	GRAMS	WEIGHT PERCENT
Polyester C Pellets	800.0	17.82
Joncryl-678 Flakes	400.0	8.91
Normal Propanol	490.0	10.91
Ammonium Hydroxide (28% Solution)	120.0	2.67
Propylene Glycol	60.0	1.34
Water	2620.0	58.35

acrylic/ polyester blend[15] was 0.127. The polymer blends were stable and did not separate into two distinct phases on storage for several months at room temperature.

The process allows the formation of polymer blends containing acrylic resin/polyester polymer at a weight ratio in the range of 10:90 to 35:65. The Brookfield viscosity of these polymer blends was about 50-500 cps, as measured at 30 rpm.

The viscosity of the 25/75 (wt/wt) ratio of acrylic resin and polyester C polymer blend, as a function of solid content is presented in Figure 2. Several batches of the polymer blends were prepared using the process described in this article. For the acrylic resin/polyester C polymer blends, with a 25/75 ratio, the viscosity remains constant up to 26 % solid content. Beyond 26 % solid content, the viscosity abruptly increases for certain polymer blends. The abrupt increase in the polymer blend viscosity occurred in the range of 26-28 % solid content for most of the batches. It appears that the process variability contributes to this variation in the viscosity. The viscosity of the polymer blends is very low (e.g about 10 cps at 30 rpm) up to 26 % solid content.

As several applications demand certain viscosity profiles, the viscosity of polymer blends at different temperatures is measured. Tables III-V represents the viscosity data for different batches at zero, room temperature and 50°C. The viscosity of various batches remained almost constant for a period of six week. A slight decrease in viscosity was observed initially. After one week, the viscosity remained almost constant for all batches. The viscosity for several batches remained below 100 cps at 30 rpm (Table III).

Table III. Viscosity of Joncryl-678/Polyester C
(25/75 wt/wt) Polymer Blends Containing
28.0 % solids at 0 °C

TIME (WEEK)	VISCOSITY OF POLYMER BLEND SAMPLES			
	(1)	(2)	(3)	(4)
0	34.1	152.0	141.0	146.0
1	22.0	42.1	58.1	47.1
2	23.0	52.1	65.1	57.1
3	31.9	62.1	86.8	81.2
4	25.0	54.1	56.1	58.1
5	29.1	47.1	48.1	-
6	33.2	71.1	63.1	-

Table IV. Viscosity of Joncryl-678/Polyester C
(25/75 wt/wt) Polymer Blends Containing
28.0 % solids at Room Temperature

TIME (WEEK)	VISCOSITY OF POLYMER BLEND SAMPLES			
	(1)	(2)	(3)	(4)
0	34.1	152.0	141.0	146.0
1	22.0	42.1	58.1	47.1
2	23.0	52.1	65.1	57.1
3	31.9	62.1	86.8	81.2
4	29.1	76.2	115.0	128.0
5	31.9	91.8	194.0	234.0

Table IV. Viscosity of Joncryl-678/Polyester C
(25/75 wt/wt) Polymer Blends Containing
28.0 % solids at 50°C

TIME (WEEK)	VISCOSITY OF POLYMER BLEND SAMPLES			
	(1)	(2)	(3)	(4)
0	34.1	166.0	154.0	174.0
1	30.1	76.8	84.2	89.2
2	44.1	75.2	41.1	71.1
3	71.1	50.1	30.1	43.1
4	132.0	40.1	28.0	35.1
5	199.0	35.1	38.1	33.2

The same trend in viscosity variation was observed at room temperature and 50°C. For certain batches, the viscosity decrease was less pronounced initially as compared to some batches. After five weeks, a slight increase in viscosity was observed for all batches at room temperature, zero and 50°C.

ACRYLIC/POLYESTER POLYMER BLENDS WITHOUT ALCOHOL

Acrylic resins belonging to this group include Joncryl-682 and Joncryl-683. It was found that a low viscosity polymer blend, in water, can be prepared with polyester C and acrylic resins (e.g. Joncryl-682 and Joncryl-683), without using volatile organic compounds (e.g. alcohols). The compatibility of the polymer blend depends on the properties of the acrylic resin and polyester polymers. The composition of the polymer blends is shown in Table VI.

Table VI. Composition of Waterborne Acrylic/Polyester Blends in the Absence of Alcohol

INGREDIENTS	AMOUNT	
	GRAMS	WEIGHT PERCENT
Polyester C Pellets	408.0	21.0
Joncryl-682 Flakes	175.0	9.0
Ammonium Hydroxide (28 5 Solution)	44.0	2.2
Water	1315.0	67.8

Several batches of acrylic/polyester blends were made with solid contents in the range of 28.0 - 35.0 wt%. The acid number of the polymer blends ranges between 50.0 - 80.0, depending on the ratio of acrylic resin to polyester polymer in the aqueous polymer blends. As the acrylic resin content in the

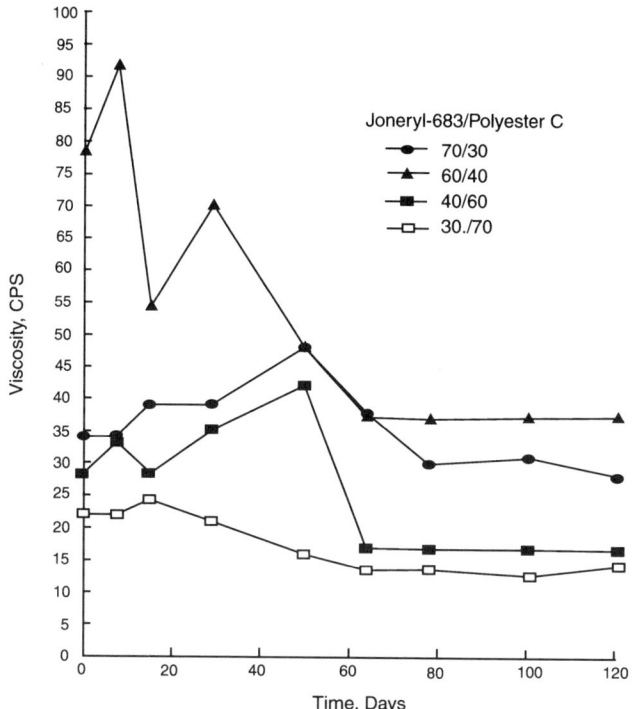

Figure 3. Viscosity as a Function of Time for
Joncryl-683/Polyester C Polymer Blends
at Zero °C.

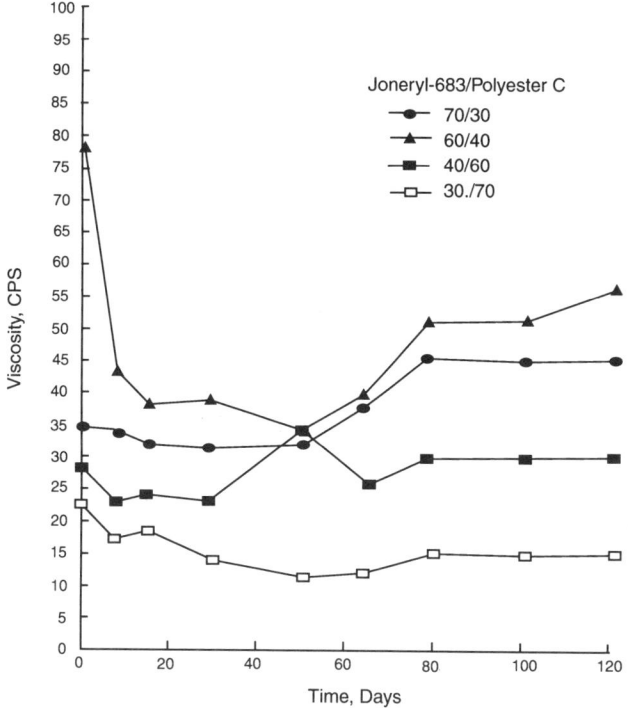

Figure 4. Viscosity as a Function of Time for
Joncryl-683/Polyester C Polymer Blends
at Room Temperature (24°C).

polymer blends increases, the acid number also increases, due to the high acid number of acrylic resin, as compared to the polyester polymer. The pH of the polymer blends were in the range of 7.15 - 8.00 for most batches.

The waterborne acrylic resin/polyester polymer blends, using Joncryl-682 and Joncryl-683, can be prepared using the present process, with an acrylic/polyester ratio that varies from 70/30-30/70 (wt/wt). The viscosity of the various batches monitored as a function of time. Results indicate that the viscosity of the waterborne polymer blends varies from 8.0-100.0 cps.

Figures 3-5 show the variation in viscosity for the waterborne Joncryl-683/Polyester C polymer blends at different temperatures. The viscosity at various Joncryl-683/Polyester C polymer ratios, as a function of time at zero °C, is presented in Figure 3. The initial viscosity for various batches was in the range of 20.0 - 80.0 cps at 30 rpm. In some cases, a slight increase in viscosity was observed in the first few days. However, an irregular variation in viscosity was observed during an initial period (e.g, up to 8 weeks). Beyond 8 weeks, the viscosity of the waterborne polymer blends remained almost constant.

The viscosity at room temperature for waterborne polymer blends as a function of time is recorded in Figure 4. At room temperature, a decrease in initial viscosity was observed with storage time. After 2 weeks, the viscosity of the waterborne polymer blends remained almost constant for up to 4-5 weeks. Beyond this time period, a slight increase in the viscosity was observed as a function of time. The viscosity of the polymer blends remains in the range of 20.0 - 80.0 cps at 30 rpm.

The viscosity of the waterborne polymer blends at 50°C is presented in Figure 5. A significant variation in the viscosity was observed at 50°C, as compared to zero °C and room temperature. The waterborne polymer blends with a high acrylic resin/polyester polymer ratio exhibited a considerable increase in viscosity (e.g. gelled) in a few weeks. The waterborne Joncryl-683/Polyester C polymer blends with 70/30 - 50/50 (wt/wt) ratios gelled, while the polymer blends with 40/60 and 30/70 ratios exhibited a decrease in the initial viscosity. After a period of 4 weeks, the viscosity of the polymer blends remained almost constant.

CONCLUSIONS

It is demonstrated that the waterborne acrylic/polyester blends can be prepared using the process described in this article. These polymer

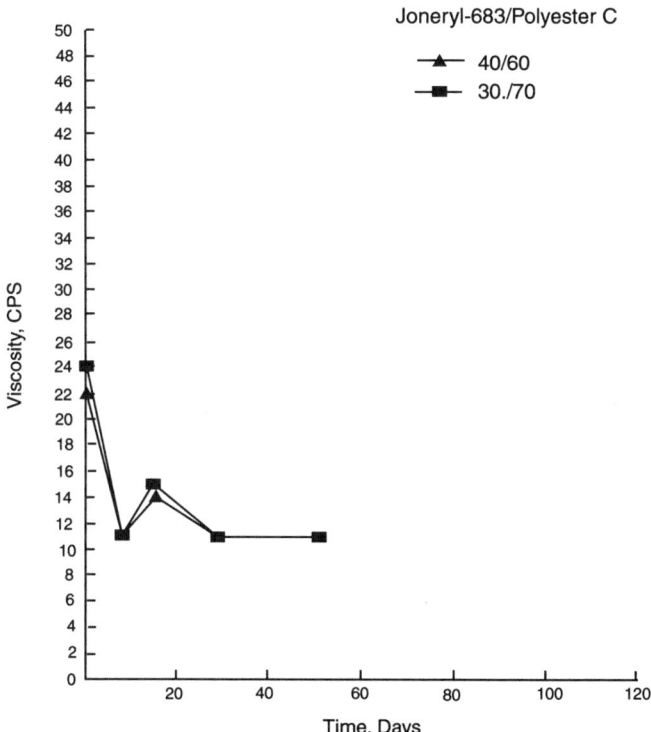

Figure 5. Viscosity as a Function of Time for Joncryl-683/Polyester C Polymer Blends at 50°C.

blends can be utilized in pigment grinding, and as a binder for various coating and printing applications. As described above, Joncryl-67, Joncryl-678 and Joncryl-680 require about 5.0 - 10.0 wt% of normal propanol or isopropanol to form stable polymer blends, while Joncryl-682 and Joncryl-683 can form stable waterborne polyester blends without alcohol. The viscosity of the waterborne polymer blends is in the range of 10.0 - 500.0 cps, at a high solids content (e.g. 25.0-35.0 wt%). Therefore, these polymer blends can be used in waterborne inks, coatings, paints and textile coating applications.

ACKNOWLEDGEMENTS

The author would like to gratefully acknowledge the management, and Eastman Chemical Company for granting permission to publish this work on the formation of waterborne acrylic/polyester polymer blends.

REFERENCES

1. Kai, S. F., Rai, D. N., Shaw, M. A., Ryan, G. L. and Collins, W.; Chemical Treatment of Overspray Paints, J. Coatings Tech., **63(789)**, 55 (1991).

2. Wagstaff, I.; Waterborne Basecoats, Proc. ESD/ASM Advanced Coating Tech. Conf., pp. 43-48, June (1991).

3. Mathur, L. K., Forbes, J. and Yelvigi, M.; Characterization Techniques for the Aqueous Film Coating Process, Pharmaceutical Technology, pp. 42-56, October (1984).

4. Fox, C. B.; Production Experience with Automotive Waterborne Coatings, Proc. ESD/ASM Advanced Coating Conf., pp. 161-166, June (1991).

5. Surface Phenomena and Additives in Water-Based Coatings and Printing Technology, M. K. Sharma (Editor), Plenum Publishing Company, New York, (1991).

6. Organic Coatings: Science and Technology, Volume 1: Film Formation, Components, and Appearance, Z. W. Wicks, Jr., F. N. Jones and S. P. Pappas (Editors), John Wiley & Sons Inc., New York, (1992).

7. Vandezande, G. A. and Rudin, A.; Novel Composite Latex Particles for Use in Coatings, J. Coatings Tech., **66(828)**, 99-108 (1994).

8. Sugama, T. and Taylor, C.; Pyrolysis-Induced Polymetallosiloxane Coatings for Aluminum Substrates, J. Mater. Sci., **27**, 1723 (1992).

9. Warminster, F. R. and Crochowski, R. J.; Acrylic Modifiers Which Impart Impact Resistance and Transparency to Vinyl Chloride Polymers, U.S. Patent No. 3,426,101 (1965).

10. Sugama, T., Carciello, N. and Taylor, C.; Pyrogenic Polygermanosiloxane Coatings for Aluminum Substrates, J. Non-Cryst. Solids, **134**, 58 (1991).

11. Wegmann, A.; Novel Waterborne Epoxy Resin Emulsion, J. Coatings Tech. **65(827)**, 27-34 (1993).

12. Lee, H. and Neville, K.; Handbook of Epoxy Resins, McGraw-Hill, New York, (1967).

13. Sugama, T., Kukacka, L. E. and Cariello, N.; Polytitanosiloxane Coatings Dervied from $Ti(OC_2H_5)_4$-Modified Organosilane Precursors, Prog. Org. Coatings, **18**, 173 (1990).

14. Sharma, M. K.; Process for Preparing Blends of Polyesters and Acrylic Polymers, U.S. Patent No. 5,218,032 (1993).

15. Sharma, M. K.; Process for Preparing Sulfo-Polyester/Acrylic Resin Blends Without Volatile Organic Compounds, U.S. Patent No. 5,294,650 (1994).

PIGMENT GRINDING USING WATERBORNE ACRYLIC

RESIN/POLYESTER POLYMER BLENDS: PART-B

Mahendra K. Sharma and Hieu D. Phan

Research Laboratories
Eastman Chemical Company
P. O. Box 1972
Kingsport, TN 37662 (USA)

ABSTRACT

Several waterborne acrylic resin/polyester polymer blends were used for grinding pigments (e.g. millbases) for different applications such as waterborne inks, coatings and paints. Commercially available pigments contained mostly acrylic resins as a grinding vehicle. These pigments usually exhibited a limited applications with polyesters, when used as a binder in waterborne coating and ink formulations. In order to form compatible millbases, the pigments were ground with acrylic/polyester blends. It was found that the pigment millbases prepared with acrylic resin/polyester blends could be used with the acrylic polymers and/or polyester polymers in formulating waterborne coatings, paints and inks. The film properties of these waterborne coatings and inks were excellent. The water resistance of the film on paper, aluminum foil and polymer substrates was superior than that of the film containing polyester alone. The blocking temperature for film containing polymer blends was 20 - 30°F higher than that of the polyester containing films.

INTRODUCTION

Organic solvents that evaporate during the formulations and applications of coatings, inks, and paints contribute significantly to a wide variety of air quality problems. Among various factors, sunlight is one of the key factors that cause

Surface Phenomena and Latexes in Waterborne Coatings and Printing Technologies, Edited by M.K. Sharma, Plenum Press, New York, 1995

57

these chemicals to react each other, resulting in the formation of different products, better known as "photochemical smog". These polluting organic solvents are most commonly referred to as Volatile Organic Compounds (VOC's). Ozone is formed by photochemical reactions between nitrogen oxides from fuel combustion and VOC's. Elevated ozone concentrations reduce lung functions, aggravate allergies, damage vegetation and cause eye irritations.

Consequently, the Environmental Protection Agency (EPA) and local Air Quality Regulators have stepped up their efforts to regulate the amount of VOC's released to atmosphere. Although pigment grinding, millbase, coating, paint and ink formulations represent only a small segment of the total market, any reduction of VOC's from these products will help achieve acceptable environmental conditions, and help prepare the industry for the probability of more stringent air pollution regulations in the near future.

In addition to the environmental and health safety concerns, the currently available pigments have a limited use in the waterborne ink and coating formulations. In order to make these pigment usable in waterborne coatings, paints and inks, the surface modification of the pigment is required. The surface modification involves wetting and dispersing phenomena.[1-9] Several additives[10] are used to disperse pigments in water for their incorporation in waterborne ink and coating formulations. These additives provide an excellent wetting and dispersion to the pigments, but their presence in the formulation usually contributes to the problems related to the processing during applications, as well as to the properties of the dried films.

The use of water-dispersible polyester polymers usually does not require any wetting and dispersing agents as these polyesters contain sulfo- groups. However, the aqueous millbases prepared using polyesters have a limited use in formulating waterborne inks and coatings, which contain acrylic polymers as a binder. If the polyesters are used as a binder in the formulations, the formulated products provide an excellent stability. In order to avoid the limited use of these millbases, it was decided to prepare pigments for waterborne inks and coatings by grinding with a blend of acrylic resin and polyester polymer. These millbases were compared with millbases containing either acrylic resin or polyester polymer alone.

The acrylic resin/polyester polymer described earlier in one of the chapters of this book, were used to prepare millbases by grinding pigments for their applications in waterborne paints, coatings and inks. The waterborne coatings, paints and inks were formulated using the millbases prepared by grinding pigments in the presence of the polymer blends. The properties of dried films were evaluated, and compared with the pigments containing polyester only instead of a blend of the acrylic/polyester polymer.

EXPERIMENTAL

MATERIALS

Several blue, green and red pigments were used for grinding to prepare waterborne millbases with the acrylic/polyester polymer blends. These pigments were purchased from the Sun Chemical Company, BASF Corporation and Ciba-Geigy Company. The waterborne acrylic/polyester polymer blends were used as a grinding vehicle. For preparing waterborne polymer blends, the acrylic resins, commercially known as "Joncryl", used were from the S. C. Johnson and Sons, Inc., while polyesters used were from the Eastman Chemical Company. The polyester used in pigment grinding is referred to as polyester C, which has glass transition temperature (T_g) about 55.0°C The distilled water was used through out the experiments.

The paint shaker was used for pigment grinding. The waterborne pigments prepared using polymer blends were incorporated to formulate coatings and inks. These inks were evaluated for the properties of films on various substrates.

METHODS

The desired pigment was added to the diluted waterborne acrylic resin/polyester polymer blend. The mixture was then shaken with glass beads for four hours on a paint shaker. The mixture was filtered through cheese cloth. The ground pigment (e.g. millbase) was stored in the plastic container. The millbase prepared in the presence of polymer blends were characterized by measuring particle size, size distribution, surface area and viscosity.

For particle size, size distribution and surface area measurements, a light scattering method was used (Microtrac Analyzer, Leeds and Northrup Instruments, St. Petersburg, Fl). The viscosity of the material was measured by Broookfield Viscometer.

These millbases were used for preparing waterborne coatings/inks. The millbases were diluted with selected waterborne polymer for preparing coatings/inks for desired color strength of the film. The film drawdowns were made on various paper, aluninum foil and polymer substrates for laboratory evaluation using different size bars/rods and RK Control Coater (Model No. KCC 101) from RK Print-Coat Instruments Limited, Litlington, England. Using the RK Coater, uniform film on selected substrates can be formed at controlled coater speed. These films were evaluated for various film properties such as water resistance, heat resistance, scuff resistance and gloss as discussed previously[11,12].

RESULTS AND DISCUSSIONS

PIGMENT GRINDING (e.g. MILLBASES)

Several pigments were ground in the presence of the acrylic/polyester C polymer blends, acrylic resin and polyester C alone, using paint shaker. The composition of the pigment dispersions (e.g. millbases) is recorded in Table I. The polymer blends contained 30.0 % solids with the Joncryl-67/polyester C in a ratio 30/70 (wt/wt) were used for pigment grinding. The solid content of water-dispersible polyester C polymer was about 30.0 %.

These millbases were characterized by measuring particle size, size distribution and surface area by light scattering method. Figures 1-4 illustrate the particle size distribution of the dispersed pigments in the presence of acrylic resin/polyester C polymer blend and polyester C polymer alone. The average pigment particle size is about 2.0 microns for green millbase, whereas the particle size for orange millbase is about 1.0 micron. Results demonstrated that the orange pigment can provide smaller particle size as compared to green pigment when polymer blend and polyester polymer were used as a grinding vehicle.

The polymer blends containing Joncryl-678/polyester C, Joncryl-683/polyester C and Joncryl-682/polyester C were also used as a grinding vehicle. The particle size, size distribution and surface area of the dispersed pigments were in the same range as observed with green and orange pigments. The particle size data at 10%, 50% and 90% sample volume for blue millbases prepared using the acrylic resin/polyester C polymer (30/70 wt/wt) blend are recorded in Table II.

The particle size at 50% of the sample (based on volume) is in the range of 2.0, - 4.0 microns, whereas the surface area is in between 2.5 - 5.5 m^2/cc for different batches of the blue millbase. For comparing millbases, the blue millbase was also prepared using Joncryl-678 resin alone. The average pigment particle size for blue millbase prepared with Joncrly-678 resin was 3.07 micron and surface area was 3.78 m^2/cc. However, the particle size distribution was bimodal with Joncryl-678 resin. The distribution curve was broad with some particle (e.g about 4.0% of the sample based on volume) with size in the range of 50.0 microns.

These blue millbases were examined for storage stability at room temperature. The viscosity of blue millbases containing polymer blends was lower as compared to millbases containing acrylic resin alone. It was found that blue millbases prepared with Joncryl-678 alone gelled in a three week period, while the millbases prepared in the presence of Joncryl-678/polyester C polymer blends did not gel during the same time period.

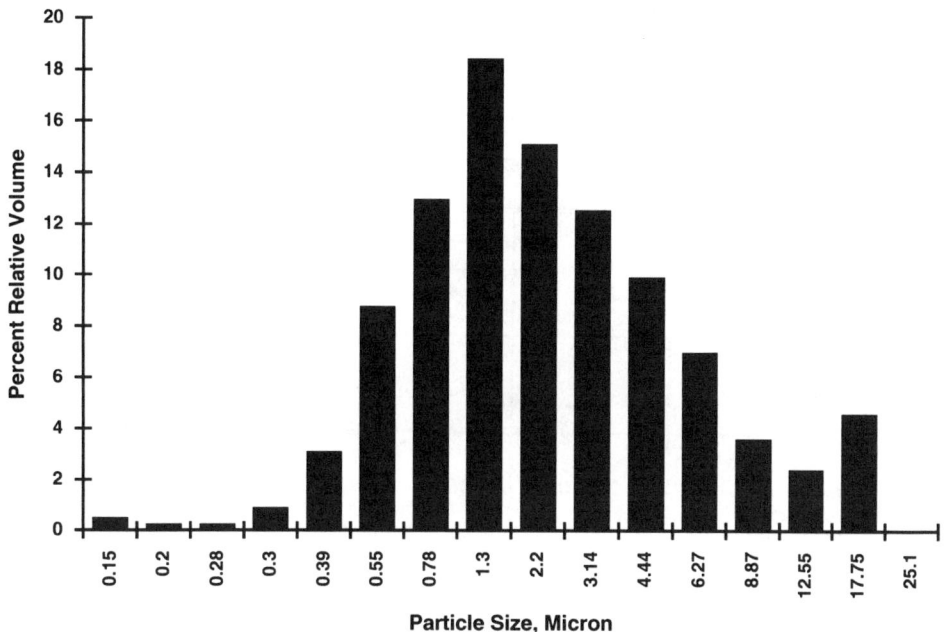

Figure 1. Particle Size of Green Millbase Containing Joncryl-678/Polyester C Polymer Blends.

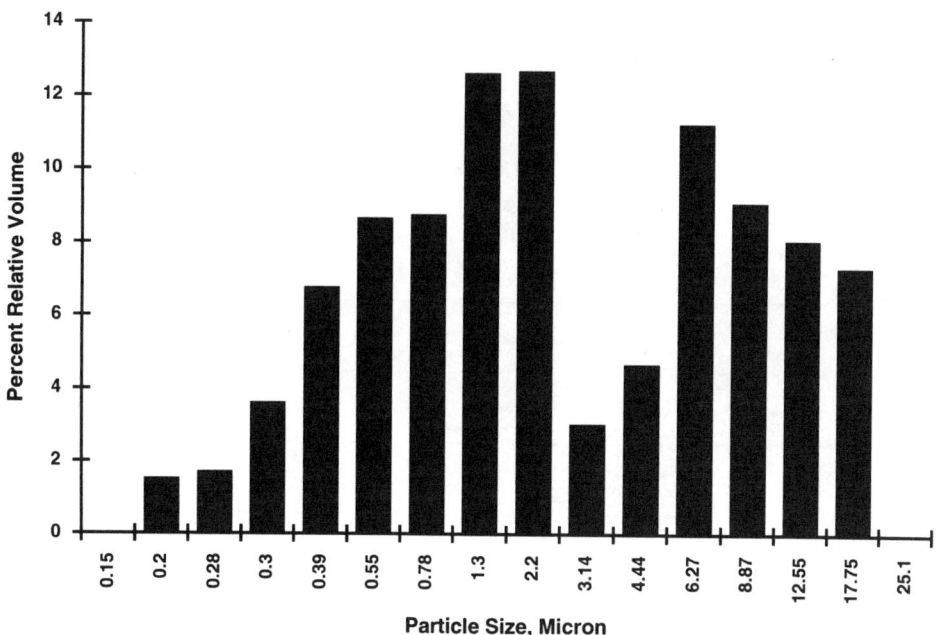

Figure 2. Particle Size of Green Millbase Containing Polyester C Polymer.

Table I. The Composition of Millbases Prepared Using
Acrylic/Polyester Polymer Blends

INGREDIENTS	JONCRYL-67/ POLYESTER C BLENDS		POLYESTER C POLYMER	
	SUNFAST GREEN 36	ORANGE 46	SUNFAST GREEN 36	ORANGE 46
Pigment (%)	50.0	50.0	50.0	50.0
Grinding Vehicle (%)	25.0	25.0	25.0	25.0
Water	25.0	25.0	25.0	25.0

Table II. Particle Size and Surface Area of Several Blue
Millbases Prepared Using Joncryl-678 Resin/
Polyester C polymer (30/70 wt/wt) Blends

BATCHES	PIGMENT PARTICLE SIZE (MICRON)			SURFACE AREA (m²/cc)
	10%	50%	90%	
1	0.48	2.03	5.64	5.12
2	0.56	3.50	8.83	3.66
3	0.61	3.27	9.17	3.35
4	0.55	3.07	9.18	3.77
5	0.84	3.75	9.85	2.90
6	0.80	3.54	8.44	3.05

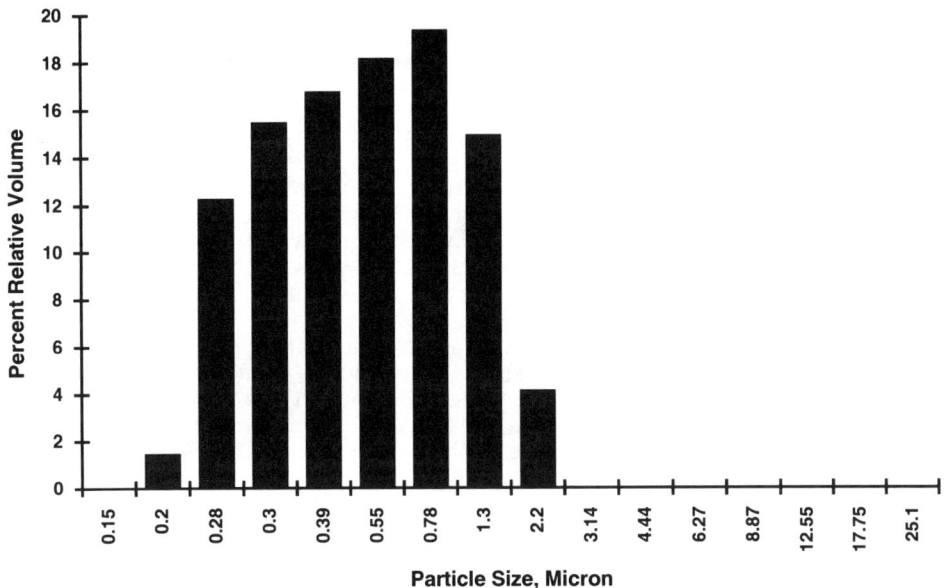

Figure 3. Particle Size of Red Millbase Containing Joncryl-678/Polyester C Polymer Blends.

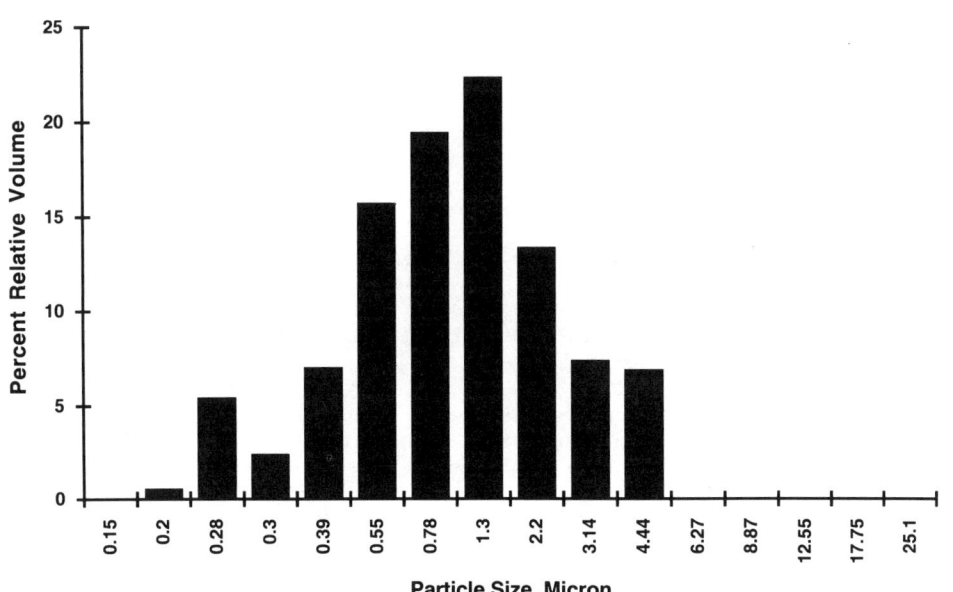

Figure 4. Particle Size of Red Millbase Containing Polyester C Polymer.

PROPERTIES OF COATING/INK FILMS CONTAINING MILLBASES

The waterborne coatings/inks were prepared using various polymers as a binder. The pigment dispersions (e.g. millbases) were diluted with selected binder during continuous stirring the mixture. The additives such as defoamer, wetting agent, levelling agent etc. were added to the formulation as required to achieve desired process and/or end-use properties of the film. As the millbases contain the acrylic/polyester polymer blends, the coatings/inks can be formulated with either acrylic polymer or polyester polymer incorporated as a binder. The millbases prepared with acrylic resin alone had limited compatibility with polyester polymers used as a binder, and therefore can not form stable coatings/inks. The composition of several coatings/inks is outlined in Table III.

The waterborne coating/inks were evaluated for properties of the film. The film was formed on paper, aluminum foil and polyester substrates with different bars. These samples were allowed to dry for 24 hours at an ambient temperature or dried in oven at 100°C for 3 seconds. These dried films were tested as follows:

WATER RESISTANCE: The water resistance of the film was determined by a water spot test. Distilled water drops were left for 5, 10, 15 and 20 minutes and then wiped off gently with a facial tissue paper. The integrity of the film was visually assessed. The water spot test is rated as follows:

1. Poor: Total film removed

2. Fair: Partial film removed

3. Good: Dull or discolor film, but no removal

4. Excellent: The film is substantially unchanged

The coatings/inks containing blue millbase and different binders were used to film drawdowns in the laboratory. Results obtained are recorded in Table IV.

The similar results were obtained on aluminum foil and polymer substrates. It appears that the water resistance is improved using the acrylic resin/polyester polymer blends either as a grinding vehicle or binder in formulating waterborne coatings/inks. However, it was found that the Joncryl-682/polyester C polymer and Joncryl-683/polyester C polymer blends based coatings/inks had inferior water resistance as compared to the inks containing Joncryl-67/polyester C and Joncryl-678/polyester C polymer blends. The adhesion of the ink on paper, aluminum foil and polymer substrates was excellent.

Table III. Composition of Waterborne Coatings/Inks Containing
Different Polymer Blends and Polymers as a Binder

INGREDIENTS	JONCRYL-67/ POLYESTER C POLYMER BLEND	JONCRYL-678/ POLYESTER C POLYMER BLEND	JONCRYL-678 RESIN	POLYESTER C POLYMER
Blue Millbase	12.5	12.3	12.6	12.5
Binder (30% Solid)	75.0	74.8	75.1	74.9
Water	12.5	12.9	12.3	12.6
Defoamer	0.10	0.10	0.10	-

Table IV: Water Resistance of the film on Paper.

BINDERS	WATER RESISTANCE RATING AT DIFFERENT TIMES (MINUTES)			
	5	10	15	20
Joncryl-67/ Polyester C Blend	4	4	4	4
Joncryl-678/ Polyester C Blend	4	4	4	4
Joncryl-678 resin	4	4	4	3.5
Polyester C Polymer	4	4	3.5	3.0

HEAT RESISTANCE (e.g. BLOCKING): The heat resistance was measured by the PI Sentinel Heat Sealer at 40 psi for 5 second. The temperature at which the film is blocked, is called blocking temperature. For heat resistance study, the samples were folded face-to-face coated/printed surface, than placed under Heat Sealer at different temperatures. The test is repeated at each temperature until the blocking occurred. The integrity of the film was rated as follows:

1. Poor: Picked and completely film removed

2. Fair: Picked, but partial film removed

3. Good: Slightly picked, but no film removed

4. Excellent: No picking and no film removed

Results obtained are presented in Table V. The films containing polymer blends exhibited higher blocking temperature as compared to the films containing polyester alone. The films containing acrylic resin/polyester polymer blends were blocked at temperature between 140-160°F, whereas the polyester based film s were blocked at about 100 -130°F.

Table V. Blocking Temperature for Blue Ink Film on Paper

GRINDING VEHICLE	BLOCKING TEMPERATURE (°F) OF FILMS CONTAINING DIFFERENT BINDERS			
	JONCRYL-67/ POLYESTER C BLEND	JONCRLY-678/ POLYESTER C BLEND	JONCRYL-678 RESIN	POLYESTER C POLYMER
JONCRYL-67/ POLYESTER C BLEND	150 - 160	140 - 150	140 - 150	120 - 130
JONCRLY-678/ POLYESTER C BLEND	140 - 150	140 - 150	140 - 150	110 - 130
JONCRYL-678 RESIN	150 - 160	140 - 150	140 - 150	110 -120
POLYESTER C POLYMER	140 - 150	140 - 150	130 - 140	100 - 110

FILM GLOSS: The gloss of the film containing polymer blends was slightly higher than that of the polyester based films. Based on the pigment content in the film, the gloss at 60 degree was in the range of 40 - 60, which was about 10 units higher than the polyester based films.

CONCLUSIONS

It has been shown that the waterborne acrylic resin/polyester polymer blends can be used to grinding pigments for their incorporation in formulating coatings, paints and inks. The viscosity of the millbases containing acrylic resin/polyester polymer blends is lower as compared to the millbases containing either acrylic resin or polyester polymer alone. Therefore, the storage stability of the millbases can be extended by using waterborne polymer blends. These millbases were used to formulate waterborne coatings/inks. The water resistance, gloss and blocking resistance of the film were superior as compared to the film containing polyester alone either as a binder or as a grinding vehicle. The blocking temperature was about 20-30°F higher for polymer blend based films than that of the films based on polyester alone. Therefore, the formulators can benefit by using the waterborne acrylic resin/polyester polymer blends instead of using polyester or acrylic resin alone in pigment grinding and coating/ink formulation work.

ACKNOWLEDGEMENT

The author would like to gratefully acknowledge the management, and the Eastman Chemical Company for granting permission to publish this work on the pigment grinding in the presence of acrylic resin/polyester polymer blends.

REFERENCES

1. Patton, C. T.; Paint Flow and Pigment Dispersions, John Wiley & Sons, Inc., New York, pp.256 (1994).

2. Sen, G.; Resins for Water-Based Inks: Color Development and Grinding, American Ink Maker, pp. 37-40, 180, December (1987).

3. Watson, W. M.; Adhesion to Polyethylene with Water-Based Inks: American Ink Maker, pp. 38-106, October (1984).

4. Micale, F. J., Iwasa, S., Lavelle, J., Sunday, S. and Fetsko, J. M.; The Role of Wetting: Part-1, Lithography; American Ink Maker, pp. 44-54, September (1989).

5. Micale, F. J., Iwasa, S., Lavelle, J., Sunday, S. and Fetsko, J. M.; The Role of Wetting: Part-2, Flexography; American Ink Maker, pp. 25-35, October (1989).

6. Menguro, K. and Esumi, K.; Interaction Between Pigment and Surfactants, Journal of Coating Tech., **62(786)**, 67-77, (1990).

7. Klarfeld, D.; Water-Based Ink Additives: From Drawing Board to Press, American Ink Maker, pp. 50-58, October (1989).

8. Domingue, J.; A Dynamic Approach to Surface Energy and Wettability Phenomenon in Flexography, IN *"Surface Phenomena and Additives in Water-Based Coatings and Printing Technology" (M. K. Sharma, Editor)*, Plenum Publishing Company, New York, pp. 163-169, (1991).

9. Bassemir, R. W. and Krishnan, R.; Practical Applications of Surface Energy Measurements in Flexography, Flexo, July (1990).

10. Vash, R.; Wetting and Dispersing, IN *"Handbook of Coatings Additives (L. J. Calbo)"*, Marcel Dekker, New York, pp. 511-539 (1987).

11. Sharma, M. K. and Phan, H. D.; Water-Based Flexo and Gravure Inks Containing Eastman AQ Polyesters, IN *"Surface Phenomena and Additives in Water-Based Coatings and Printing Technology" (M. K. Sharma, Editor)*, Plenum Publishing Company, New York, pp. 27-41, (1991).

12. Sharma, M. K. and Phan, H. D.; Ink Millbase and Method for Preparation Thereof, U.S. Patent No. 5,162,399 (1992).

THE DISPERSION OF ORGANIC PIGMENTS IN

AQUEOUS MEDIA

G. V. Calder and D. M. Wilson

S. C. Johnson Polymers
1525 Howe Street
Racine, WI 53403

ABSTRACT

The dispersion of colored organic pigments in aqueous media is important in graphic arts, paints and coatings. Restraints on the release of volatile organic compounds (VOC's) into the atmosphere have compelled these industries to adopt waterborne vehicles. In addition existing or pending regulations on the use of, and disposal of pigmented products containing heavy metals can require the use of more expensive organic pigments. Consequently, pigment dispersant manufacturers must refine the polymers used to disperse colored, organic pigments to obtain higher efficiency and performance. Therefore, the design of polymers as a grinding vehicle specially for organic pigments is presented. Results demonstrate that the waterborne organic pigment dispersions can be prepared for high quality printing and coating formulations.

INTRODUCTION

Monomeric surfactants and inorganic polyelectrolytes, such as polyphosphates, can easily disperse hydrophilic, inorganic, mineral pigments. However, polymeric, organic, polyelectrolytes, such as styrene/acrylic acid copolymers, have significant advantages in the dispersion of hydrophobic,

Surface Phenomena and Latexes in Waterborne Coatings and Printing Technologies, Edited by M.K. Sharma, Plenum Press, New York, 1995

71

colored, organic pigments. In addition, municipal, state, and federal regulations regarding the use and disposal of pigments containing heavy metals such as Hg, Pb, Cd, Zn, Cr(VI), and possibly Ba, are becoming increasingly restrictive. As a result, in many applications, organic pigments are replacing inorganic pigments containing heavy metals. The organic pigments are more expensive than the heavy metal inorganic pigments. The price of several organic and inorganic pigments are listed in Table I. Due to increasing demand in market place to meet current environmental regulations, the maximum dispersion efficiency of these more expensive organic pigments becomes increasing important.

Obviously, maximum utilization of these expensive organic pigments is economically important.

PIGMENT DISPERSION

Historically, the evaluation of pigment dispersion has been largely empirical, relying on the keen eye of an experienced formulator. The classic work of Kubelka and Munk[1] in 1931, and various subsequent refinements have laid the foundation for the quantitative, objective assessment of color. Previous work[2-5] provides an up to date treatment of the objective assessment of color. In this work we attempted to use objective, quantitative criteria for dispersion in so far as possible. The goal of this work is to design new polymeric dispersants for organic pigments with dispersion properties superior to industry standards.

Table I. Typical Prices of Organic and Inorganic Pigments (American Paint & Coatings, March 2, 1992).

INORGANIC PIGMENTS	PRICE ($/LB)	ORGANIC PIGMENTS	PRICE ($/LB)
Yellow Pigment (Lead Chromate)	1.05	Yellow Pigment (Hansa Yellow)	6.50
Orange Pigment (Lead Chromate)	1.20	Orange Pigment (Benzidine Orange)	6.25
Red Pigment (Cadmium Mercury Red)	8.25	Red Pigment (Quinacridone)	30.00

The quality of a pigment dispersion is measured by several, sometimes conflicting and confounded parameters:

Tinctorial Strength

Gloss

Transparency

Rheology

Stability

The first step in the process of pigment dispersion is the milling of the pigment. In this step, air adsorbed on the pigment particles is replaced by the fluid media, and pigment agglomerates are broken up by mechanical action. Traditionally, the milling process has been carried out with:

Ball Mills

Sand Mills

High Speed Mixers

Each of these milling methods has advantages and limitations.

What is desired of a pigment mill in a research laboratory context is:

Small Sample Size

Rapid Dispersion

Good Temperature Control

Easy Cleanup

Correlation With Production Equipment

EXPERIMENTAL

After considerable searching, we found a media mill manufactured by Eiger Machinery that provides a good compromise of these objectives. This is not intended to be

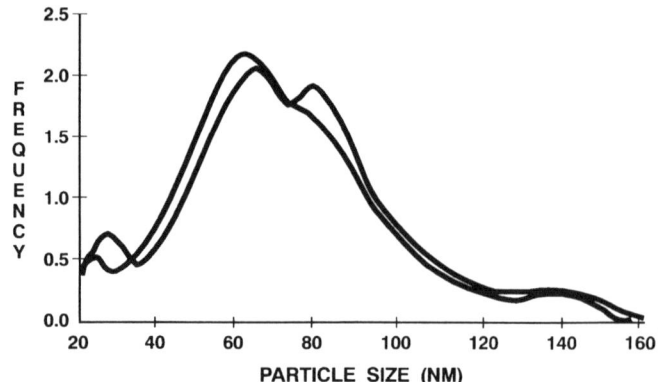

Figure 1. Particle Size of the Dispersions Prepared in Laboratory as compared to the Commercially Prepared Dispersions.

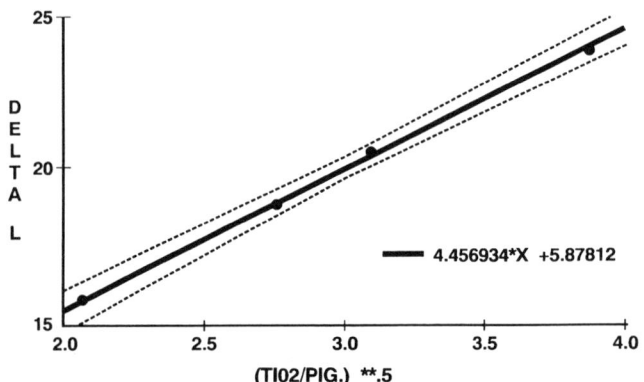

Figure 2. A Correlation Test Between $|DELTA\ L^*|$ and $(TiO_2/PIG.)**.5$ for Bleach/Blue Table.

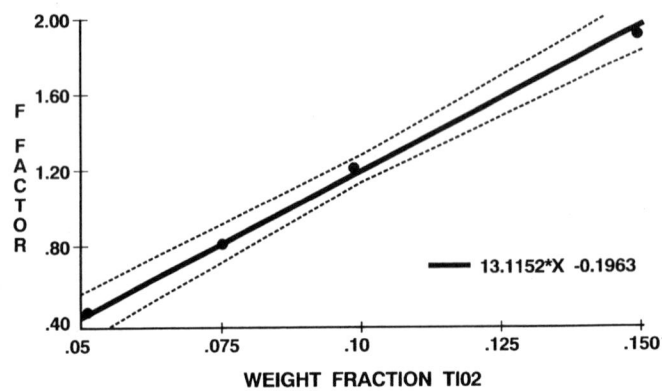

Figure 3. Kubelka-Monk Analysis of Phthalo Blue 15:3 Data.

an endorsement for that particular equipment, but simply a statement of our evaluation results versus other milling methods.

SAMPLE SIZE

A small size of the sample can be prepared. Samples of 75 to 300 gm. can be processed.

FAST DISPERSION

Fifteen minutes of the continuously recycled dispersion formulation gives dispersion quality comparable to twenty four hours of ball milling.

TEMPERATURE CONTROL

The mill is thermostated, and the temperature of the grind can be measured directly. Water coolers are commercially available where seasonal temperature changes influence the temperature of the cooling water.

EASY CLEANUP

The mill can be cleaned simply by recycling unpigmented vehicle or using a commercial hard surface cleaner, for pigments of similar color; for pigments of very different color, the mill can be opened easily and thoroughly cleaned by soaking the parts in an appropriate cleaning solution.

CORRELATION WITH PRODUCTION EQUIPMENT

Our experience has been that dispersions produced with this equipment correlate well with dispersion methods used industrially. Shown in Figure 1 is the particle size analysis of a dispersion compared to a commercially prepared dispersion.

PARTICLE SIZE

The size of the pigment particles, and their distribution, have an important impact on the tinctorial properties of a pigment dispersion. Hence its determination is of the utmost importance in assessing the efficiency of a dispersant polymer. Historically, the grind gauge has been the laboratory method for determining the particle size pigment dispersions; however, for most organic pigments, this test is far too crude to be of much value.

Laser light scattering can provide a measure of particle size. But because the method is based on the average cross sectional area of the scattering particles, and certain other approximations in the analysis of the data, the average particle diameter computer is suspect for a heterodisperse, dispersion of non-spherical particles.

Optical or electron microscopy,, provides another method for particle size analysis. But these techniques require careful, tedious sample preparation, and sometimes labor intensive image analysis of the results.

Capillary hydrodynamic fractionation (CHDF) offers yet another technique for determining both the average particle size of pigment dispersions, as well as the particle size distribution. Comparisons of the distributions measured by this technique compare favorably with distribution determinations measured by the more tedious microscopy techniques. In addition, CHDF provides a detailed analysis of the particle size distribution (Figure 5).

THEORETICAL CONSIDERATIONS

QUANTITATIVE EVALUATION OF COLOR

Over the past few decades, various instruments relating the visual perception of color to spectroscopic measurements have been developed. One of the more commonly used systems is the Hunter L^*, a^*, b^* color coordinate system in which the color of a surface is defined in terms of L^*, a measure of the reflectance of the film, a^*, a measure of red/green axis, and b^* a measure of the blue/yellow axis. Thus the color of a surface is defined in terms of a vector of complementary colors (ΔL^*, Δa^*, Δb^* compared to some standard. This can also be expressed in terms of a root mean square difference:

$$\Delta E = (\Delta L^{*2} + \Delta a^{*2} + \Delta b^{*2})^{1/2}$$

This difference, ΔE, only gives the difference from the standard, not how the color differs spectrally from the standard. In the Hunter model, the value of L* is defined:

$$L^* = 10.0 \ ^*(Y)^{1/2}$$

where Y is the reflectance of the surface. The reflectance of a surface is defined in terms of the index of refraction of the surface, n_2, and the index of refraction of the surrounding medium, n_1:

$$Y = ((n_2 - n_1)/(n_2 + n_1))^2$$

For air n_1 - 1.0.

For a surface whose reflectance is dominated by a mixture of two pigments, A and B, the index of refraction is given by:

$$n_{AB} = C_A \ ^* \ n_A + (1-C_A) ^*n_B$$

where C_A, n_A, and n_B are the weight fraction of pigment A, the index of refraction of pigment A, and the index of refraction of pigment B, respectively.

Thus L* should be proportional to $C_A^{1/2}$.

This model, Fresnel's Law, is observed in bleachouts of phthalo blue 15:3 with TiO_2 at dilutions varying from about 5/1 to 15/1 (Figure 2). Note the narrow 95% confidence interval, and the low residuals of the observed and calculated values of ΔL* (Table II):

Table II. Residuals of Observed and Calculated Values of L* For Phthalo Blue 15:3 and TiO_2 Bleachouts.

TiO2/Blue	ΔL* (obs.)	ΔL* (calc.)	Residual
5.0	15.82	15.84	-0.02
7.5	18.07	18.08	-0.01
10.0	20.05	19.97	+0.08
15.0	23.10	23.14	-0.04

An alternative analysis, using a simplified form of the Kubelka - Munk equations:

$$F = K/S = (1-Y)^2 / 2Y$$

yields a similar accurate fit to the data as illustrated in Figure 3.

In many cases where the pigment is a primary color -- red, blue, or yellow -- the value of the color axes ΔL*, Δa*, and Δb* is dominated by a single axis. This is illustrated in Table III.

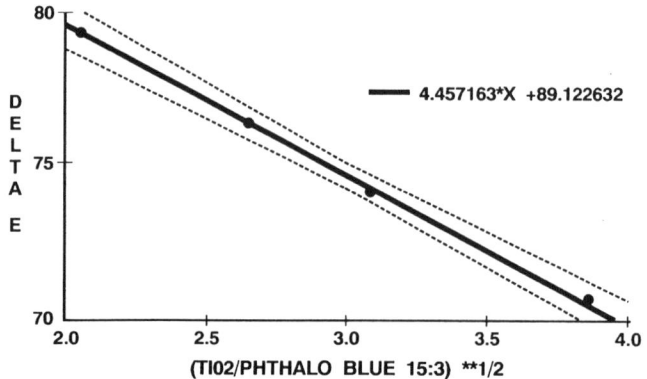

Figure 4. Delta E Versus $(TiO_2/Phthalo\ Blue\ 15:3)^{1/2}$ at Constant Color Pigment Content.

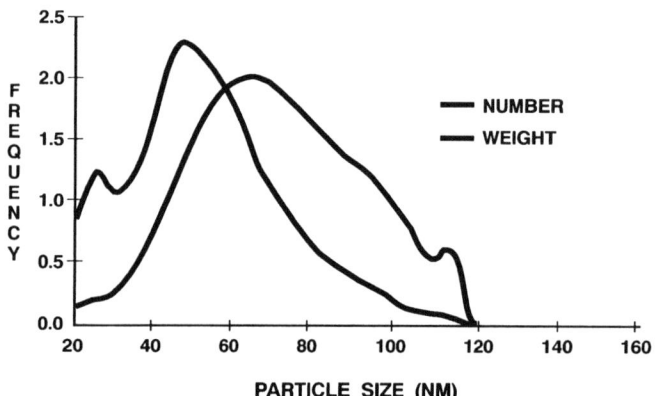

Figure 5. Particle Size Distribution of Dispersions Using Capillary Hydrodynamic Fractionation (CHDF).

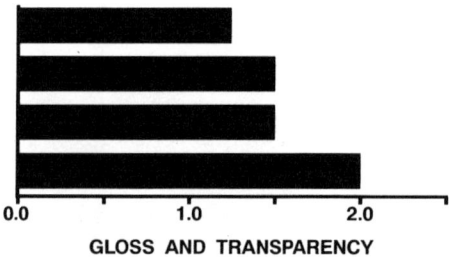

Figure 6. Relative Performance of Several Compositions of the Dispersed Resins.

Table III. Delta L*, Delta a*, and Delta b* for Bleachouts of Phthalo Blue 15:3 with TiO$_2$

TiO2/Blue	ΔL*	Δa*	Δb*
5.0	15.82	4.97	-46.53
7.5	18.07	2.83	-47.43
10.0	20.05	1.53	-47.63
15.0	23.10	0.77	-47.13

Several trends are evident. First, the absolute value of ΔE*:

$$\Delta E* = ((\Delta L*)^2 + (\Delta a*)^2 + (\Delta b*)^2)^{1/2}$$

is dominated by the value of the blue/yellow axis, Δb*; however, this term is essentially constant having a mean value of 47.18, with a standard deviation of only 0.48 over the range of dilutions with TiO$_2$. Second, the red/green axis values are small and hence do not contribute significantly to the value of ΔE*. Hence, the dependence of ΔE on the degree of bleachout with TiO$_2$ is controlled by the value of the reflectance, ΔL*. Thus, like ΔL*, ΔE* is proportional to $(C_{TiO2})^{1/2}$. This is shown in Figure 4.

RESULTS AND DISCUSSION

DESIGN OF RESIN FOR DISPERSION OF THE ORGANIC PIGMENTS

Styrene/acrylic acid copolymers for the dispersion of colored pigments in aqueous media have been an industry standard for many years. They were a spin off of floor finish technology in the early 1970's. Natural products, like shellac and alkali soluble rosin, were being replaced as leveling agents in floor finishes by synthetic styrene/acrylic acid copolymers. The natural products were also used to disperse pigments in corrugated board inks. The assumption was that the synthetic polymers could be "plugged in" to replace the natural products in the ink applications. The assumption was correct and the industry trend was to replace the natural products with the more controllable synthetic substitutes. Because these polymers were "good enough" for the traditional pigments, not much attention was paid to optimizing their composition for dispersion. However, new issues arose in the 1980's that required re-examination of the polymer compositions.

As a result of solvent emission concerns, waterborne inks were being examined for higher quality printing applications, such as the printing of film and foil, where appearance properties, ink rheology and pigment costs were more critical than in the printing of low quality paper and corrugated board. More recently, environmental concerns on the use and disposal of pigments containing heavy metals altered the selection of pigments used in the coatings and graphic arts industries. Although, more expensive, organic colored pigments are increasingly preferred over inorganic colored pigments containing Pb, Cd, Zn, Cr(VI), and possibly Ba. These trends require refinement, and in some cases redefinition, of the composition of pigment dispersing polymers for the "Environmental Pigments of the 1990's" to maximize the dispersion efficiency and properties of the more expensive organic pigments[6].

The dispersion properties of a polymer are determined by:

Type of Functionalities

Level of Functionalities

Molecular Weight

Molecular Weight Distribution

These factors control both the efficiency of dispersion and the equally important rheology of the dispersion.

Keep in mind that pigments are formulated products, so simply knowing the molecular structure of the chromophore may not provide a clear picture of how the pigment will behave during the dispersion process. Various additives and surface treatments are used in the pigment manufacturing process.

Styrene/acrylic acid copolymers are common polymers used for dispersing organic pigments. These monomers are chosen for the attributes each monomer brings to the copolymer:

Styrene

- o   high Tg for block resistance

- o   hydrophobicity for water and alkali resistance

- o   high index of refraction for high gloss

<u>Acrylic Acid</u>

- o   high Tg for block resistance

- o   hydrophilicity to balance the hydrophobe

- o   weak acid which gives up the volatile

     counterion easily

Typically the ratio of styrene to acrylic acid has been in the range of 75/25, although there is some variability in this composition. This ratio provides the level of acid functionality necessary for ease of solution in aqueous alkali or amine and resolubility in the printing process, with minimum compromise of the water and alkali resistance in the finished product.

Since the reactivity ratios of both styrene and acrylic acid are both less than 0.3, there is a high tendency for the two monomers to alternate. Thus, the monomers are approximately evenly distributed along the backbone of the polymer.

The molecular weight (weight average) has typically been in the range of 8000 to 10,000. Very lower molecular weight reduces the surface activity, and pigment wetting properties of the polymer, and stabilizing properties of the copolymer. Very higher molecular weigh tends to raise the viscosity of neutralized solutions of the polymers such that they become difficult to process, and interfere with the pigments grinding efficiency. Therefore, a balance of molecular weight and functionality, in this example acrylic acid, must be struck. A comparison of the polymer properties is shown in Table IV.

Table IV. A Comparison of the Properties of Improved
and Control Polymer

POLYMER PROPERTIES	IMPROVED POLYMER	CONTROL POLYMER
M_w	16,000	10,000
ACID Number	240	190
T_g	103	70

The relative performance of several compositions is shown in Figure 6.

CONCLUSIONS

In summary, the increased use of waterborne pigment dispersions in high quality printing and coatings applications because of regulations regarding the emission of organic solvents, and the replacement of heavy metal containing inorganic pigments with more expensive organic pigments have combined to force manufacturers of polymers for the dispersion of organic pigments to refine, or redefine, the polymer composition of those polymers to meet these new needs. Concurrently, objective, instrumental methods of the evaluation of color must replace the traditional subjective evaluation.

REFERENCES

1. Kubelka, V.P., Munk, F., Zeit, fur Tech. Physik, NrIIa, 595, (1931).

2. Billmeyer, F.W. and Salzman, M., Principles of Color Technology, ISBN 0-471-03052-X, Wiley, (1981).

3. Zollinger, H., Color Chemistry, ISBN 1-56081-149-8, VCH, (1991).

4. Wyszecki, G. and Stiles, W.S., Color Science: Concepts and Methods, Quantitative Data and Formulation, 2nd Ed., ISBN 0-471-02106-7, Wiley, (1982).

5. Judd, D.B. and Wyszecki, G., Color in Business, Science, and Industry, ISBN 0-471-45212-2, Wiley, (1975).

6. Environmental Pigments '92 (Symposium) sponsored by Falmouth Associates, Inc., Falmouth, Maine, March 3-4, (1992).

COLLOIDAL ASPECTS OF COLORANTS IN RELATION TO

AQUEOUS PRINTING

George F. Sonn

Tritech, Inc.
P. O. Box 1526
Burlington, NC 27216

ABSTRACT

Aqueous printing inks can be presented as a model colloidal system that encompasses many classical colloidal concepts. Organic pigments which comprise the bulk of the pigments utilized in aqueous printing inks are composed of large surface area. For an aqueous system, the surface is much more active than the same system in an organic solvent media. Organic pigments, to be useful as coloring agents, must be in a dispersed form in the final ink. Dispersion involves the dual mechanics of surface wetting and particle size reduction. Surface wetting can be affected by both pigment pretreatments during the manufacture of the pigment and the post treatments of surface by addition of the agents.

INTRODUCTION

The use of aqueous printing ink is accelerating in the United States due to the environmental concerns related to VOC's. Organic pigments represent a major use area in the market place. Heavy metal pigments have been virtually eliminated in an aqueous system. Several articles on pigment characterizations, properties and applications are available in the literature[1-10]. Colloidal science can describe many of the phenomena that take place in the preparation of aqueous printing inks. The aqueous system is a very active one in

Surface Phenomena and Latexes in Waterborne Coatings and Printing Technologies, Edited by M.K. Sharma, Plenum Press, New York, 1995

83

terms of colloidal activity. Organic pigments represent large surface areas and each organic pigment has specific charge parameters. Aqueous systems are dynamic and are therefore much more vulnerable than solvent systems to the effects of double layer phenomena and ionic charges. Pigments for aqueous systems must therefore be formulated with their stabilization requirements in mind. Organic pigments are prepared via a coupling process of organic intermediates. The final pigment in many cases is given end treatments for particle surface and particle size alterations. These treatments in the case of aqueous printing inks are designed to increase the acceptance of the organic pigment surface to the aqueous polymer system. The matching of these interfacial energies are based upon the principles of colloidal science. Classical colloidal solutions are stabilized by an electric double layer around the colloidal particles. Deflocculation is determined by the variation in potential energy between two approaching particles. Attraction and repulsion forces are in effect at the same time: London Van der Waals, forces attract the particles, the double layer repels them. Aqueous systems are dynamic and, therefore, much more vulnerable than solvent systems to the effects of double layer phenomena and ionic charges. The use of parameter profile to predict the use of organic pigments in an aqueous printing ink system has vast utility potential. Each organic pigment class has specific surface characteristics that are indigenous to the pigment class. The major pigment classes used in process aqueous printing inks are naphthols, rubines, phthalocyanines, rhodamines, methyl violets and diarylides. Certain exotic organic pigments such as perylenes, quinacridones, and carbazoles are utilized where high performance fastness is required.

Dispersion is essentially produced by both physico-chemical forces and mechanical forces supplied by dispersion equipment. The proper selection of wetting and dispersing agents can be reduced to a parameter system by analysis of both the vehicle surface and the pigment surface. The wetting forces inherent in the vehicle/pigment systems are well known. Due to polar nature of water, it has high electrical activity. In many cases the vehicles chosen for aqueous inks also have highly polar surfaces. The key to the pigment introduction for the system is that the pigment must match the system of the resin. The wetting agent or dispersing agent to be utilized, must aid in matching the two systems and promote dispersion. The ultimate in pigment dispersion is a particle size that yields the maximum color value of the pigment. The pigment ancillary factors of the transparency and gloss are also directly proportional to the pigment particle size. Once the ultimate particle size is obtained, the forces that promote flocculation must be balanced. Such forces include Van der Waals forces, electrokinetic potential, zeta potential and steric forces.

Instruments such as the microscope, particle size counters, electrophoretic equipment and sedimentation equipment are utilized to measure dispersion stability. Color measurement equipment, such as spectrophotometers are utilized to evaluate both color consistency and ultimate color value.

Rheology is also a key determinant of colloidal stability and affects ink performance. The preparation of a final model system for each vehicle system promotes a final optimized color dispersion system for aqueous inks. The future of aqueous inks is the further refinement of this model system, to both predict and implement ideal characteristics of the dispersions. The model discussed will be illustrated with examples in this article. Aqueous inks are an environmentally friendly avenue to all inks in the future.

Zeta potential is another measure of colloidal stability. Salts which are present in many phases of pigment manufacturing, interact proportionally with the power of the zeta potential. One of the most important factors which determine the color properties of organic pigments are particle size and size distribution. Interrelated with these is the factor of flocculation which assumes that pigment particles come together due to electrical forces between the ultimate pigment particles. Flocculation affects color value, rheology, and gloss and transparency. Color strength, transparency, and gloss generally increase with a decrease in particle size. Light fastness increases with an increase in particle size. The particle size of organic pigments are normally in the sub micron range.

Particle size measurement is therefore important, especially on the aqueous system which is being evaluated. The methods used to evaluate particle size are electron microscopy (scanning and transmission), light scattering, instruments surface area analyzers and x-ray diffraction analysis. Electron microscopy and light scattering are direct particle size measuring techniques, while surface area and x-ray are indirect techniques.

The polymers that are currently in use in aqueous printing are acrylic, shellac, polyamides, polyurethanes, oxidizable alkyds, and epoxyesters and sulfonated polyesters.

In addition to preparing an organic pigment for aqueous ink systems, the organic pigment can be prepared to match the aqueous resin system. This can be done, in some cases, by actually coating the organic pigment with the vehicle system or the use of proper surfactants .

Here again, a colloidal model can be used to help predict the outcome on the final aqueous ink system. The aqueous pigment stability, in terms of the rheology parameter, manifests itself in a pigment which does not gel, has minimal flocculate and does not separate overtime. The rheology of the systems increases with particle size, but this is also a function of pigment content.

EXPERIMENTAL

Several organic pigments with and without treatment were used to study the properties of the pigments. The pigments evaluated were treated with the amount of agent to fully cover the pigment surface area, but not have an excess of free agent to interfere with the ink system. This treatment system falls within the area of new pigment technologies.

THEORY

The key to this article is that new pigment technologies must be utilized to produce organic pigments for aqueous printing systems. This technology must be based on the colloidal premises previously discussed. The other critical feature is the degree of the fineness of the pigment achieved during the dispersion process. Pigments are dispersed from two physical forms which are presscakes or dry powders. Presscakes have average smaller particle size due to the fact that they have not been agglomerated. Specific changes to the pigment's physical properties can facilitate dispersibility and improve rheological and pigmentation characteristic in aqueous printing inks. Treated pigments and highly concentrated presscakes are the most easily dispersible products. Color chips produced from a two roll mill process also produce a very high degree of dispersion due to the intense mechanical shear involved. The use of colorants in aqueous printing is an expanding area and is geared to face the environmental challenge.

RESULTS AND DISCUSSION

The use of colorants in aqueous printing evolves into various selective processes for pigments. Pigment properties with and without treatment are recorded in Tables I and II. The treated pigment systems exhibit superior strength as compared to untreated pigment systems (Tables I and II).

Pigment manufacturers can specifically design pigments for aqueous printing inks based on customer demands. One class of pigments as an example is that of the naphthol pigments. These pigments have good alkali fastness which is necessary because many of the aqueous systems are in the pH range of 8.5 and above. These naphthols also have a wide range of bright clean colors and do not possess specific heavy metal ions. The rheology of these pigments are also acceptable as print rheology is most important in proper aqueous printing. Higher quality aqueous printing also demands higher gloss and transparency in the final print phase. Naphthol chemistry is being altered to address these needs. Also of importance is the development of a clean yellow shade naphthol. This will used as a Red Lake C and Barium 2B replacement. Continual monitoring is being done on all organic pigments to reduce to ppm the trace contaminants such as low level heavy metals. The aqueous development of organic colorants is based upon the study of interfacial phenomena. When pigments are struck and are consequently given final surface treatment or particle size alteration, such treatment is governed by interfacial phenomena. In a colloidal system, the surface is enormous compared to the mass of substance. Consequently the magnitude of surface forces such as adsorption and surface tension

Table I. Spectrophotometric Data for Phthalocyanine
        Beta Blue.

PIGMENT TREATMENT	PARAMETERS	
	STRENGTH (%)	COLOR YIELD (%)
SOLSPERSE 2700	100	20
NO TREATMENT	80	-
JONCYL LV61	100	16
NO TREATMENT	84	-

Table II. Spectrophotometric Data for Lithol Rubine.

PIGMENT TREATMENT	PARAMETERS	
	STRENGTH (%)	COLOR YIELD (%)
SOLSPERSE 2700	100	18
NO TREATMENT	82	-
JONCYL LV61	100	16
NO TREATMENT	85	-

becomes extremely large and their effect is erotical. The primary pigment particle is built up by crystal growth and accretion. In actual use the ultimate pigment particle represents a formation of clusters of particles. This phenomenon is termed flocculation and its occurrence is determined by the degree to which primary particles are kept separate. These phenomena affect transparency, gloss and color value.

The pigment stability is affected by an electrical double layer around the dispersed particles. The inner of the two layers is caused by the preferential adsorption of one species of ion at the surface of the particle. When two negatively charges clouds begin to interact and give rise to a repulsive force, which is dependent upon the distance between them. The double layer is one of the stabilizing factors in colloidal systems. In some cases, stabilization may be the result of the steric nature of the adsorbed layer. One way of improving colloidal stability is to chemically modify the pigment surface so as to take advantage of the steric nature of the adsorbed layer.

CONCLUSIONS

A surface parameter profile can be used to predict surface activity of pigments in the aqueous phase. Each pigment has its own electrical charge profile which can be correlated to the clear polymer system. Pigments are produced by mechanical attrition of crude material or by actual chemical synthesis. Both methods include salts that create high electrical charges that are subsequently washed from the pigment surface. Once the pigment is in the fluid form, it can be left as is or treated with surface active agents or various resins that alter its surface charge.

Flocculation is favored when the solution is such that the electrophoretic mobility is reduced to a small value. There is a critical value of the zeta potential which, when reached, flocculation occurs. The electrical charge and the potential resulting from its presence confer stability on lipophobic sols which prevents the particles from coming sufficiently close for the Van der Waals forces of attraction to cause flocculation. Experiments have been conducted to measure attractive forces between plane surfaces at distances up to 10,000 angrstroms. Various surface chemical groups are extremely important to the stability of colloidal systems. By varying these surface groupings and characterizing the pigment and resin interfaces, a stable colloidal system can be designed. While in many cases a relatively clean pigment may be altered where necessary to confirm to suitable charge profiles via in-situ additives, surfactants or other pigment treating agents. An empirical zeta potential can readily be measured by electrophoresis of the dispersed phase. When the double layer is compressed, the incidence of flocculation is increased. In the case of the moving particle, the effective charge can be calculated. This charge represents the net charge on the surface and is the algebraic sum of the fixed charges on the surface and the strongly adherent counter ions. Other empirical methods for measurement of electrokinetic phenomena include measurement by microscope and by sedimentation. The electrophoretic mobility of the particle can be viewed via microscopy in a circular tube apparatus. Sedimentation analysis involves the measurements of the difference of the potential gradient between the electrodes at the top and the bottom of the sedimentation vessel. In addition, pigment surfaces can be designed to modify to fit the colloidal charge profile of the resin, the more ideal will be the color value and degree of dispersion of the final color concentrate. Aquaflo has been designed based on these concepts.

A theoretical statistical model can be developed with pigments in an alkaline medium. The concept of waterborne printing and waterborne coating is relatively new, but the basic parameters of formulation and application can be more accurately qualified by the use of theoretical models. In particular, the pigment chemist must adjust to designing pigment for the aqueous system with full cognizance of the colloidal science that is involved. It cannot be too strongly emphasized that the ability to provide better pigments for the aqueous system lies in the use of a theoretical statistical model utilizing the effects of the polar characteristics of the aqueous systems.

As to environmental area, organic pigments are continually being monitored to reduce the extremely low levels of heavy metals and other trace contaminants. Aqueous printing obviously substantially reduces the VOC content during the printing process.

This trend is accelerating in the U. S. and Europe and will further reduce the amount of solvents used in coatings, paints and printing processes. Printing has dramatically improved in the aqueous ink area as to both speed and quality of print. Research and development in both aqueous printing and organic pigments for aqueous printing far out paces that of conventional ink developments.

The goal for the future is a colorant system which is specifically designed for aqueous printing inks.

REFERENCES

1. Simpson, L.A., Factors Controlling Gloss of Paint Films, Pro. Organic Coatings, **6(1)**, 1-30, (1978).

2. Kawabata, A., Effects of Pigments on the Gloss of Paint Films: Part-I, J. Japan Soc. Color Mat.; **41(2)**, 2-13, (1968).

3. Kawabata, A., Effects of Pigments on the Gloss of Paint Films: Part-II, J. Japan Soc. Color Mat.; **41(2)**, 14-25, (1968).

4. Kawabata, A., Effects of Pigments on the Gloss of Paint Films: Part-III, J. Japan Soc. Color Mat.; **41(3)**, 2-8, (1969).

5. Vash, R., Pigment Wetting and Dispersiong Additives for Water-Based Coatings and Inks, In "Surface Phenomena and Additives in Water-Based Coatings and Printing Technology" (Editor, M. K. Sharma), Plenum Publishing Corporation, New York, pp.139-149, (1991).

6. Tatman, C.C., Titanium Dioxide Particle Size Control for Designed Performance in Waterborne Coating Systems, In "Surface Phenomena and Additives in Water-Based Coatings and Printing Technology" (Editor, M. K. Sharma), Plenum Publishing Corporation, New York, pp.105-137, (1991).

7. Kern, G.M., Micale, F.J., Valenzuela, D.P. and Lavelle, J.S., Hiding Power of Aluminium Pigments in Printing Ink Films, In "Surface Phenomena and Fine Particles in Water-Based Coatings and Printing Technology" (Editors, M. K. Sharma and F. J. Micale), Plenum Publishing Corporation, New York, pp.59-69, (1991).

8. Braun, J.H., Gloss of Paint Films and the Mechanism of Pigment Involvement, J. Coating Tech., **63(799)**, 43, (1991).

9. Zorll, U., New Aspects of Gloss of Paint Film and Its Measurements, Prog. Organic Coatings, **1**, 113-155, (1972).

10. Braun, J.H. and Fields, D.P., Gloss of Paint Films: II. Effects of Pigment Size, J. Coatings Tech., **66(828)**, 93-98, (1994).

CHARACTERIZING MECHANICAL PROPERTIES OF LATEX FILM AND COATING LAYER IN PAPER COATING USING DYNAMIC MECHANICAL THERMAL ANALYZER (DMTA)

Osamu Ishikawa, Takanori Yamashita and Akira Tsuji

Emulsion Laboratory, Technical Center
Japan Synthetic Rubber Company, Ltd.
100, Kawajiri-cho, Yokkaichi,
Mie, 510, Japan

ABSTRACT

In paper coating, the pigment volume content of the coated layer is over critical pigment volume concentration (CPVC) and latex polymer used as main binder forms a discontinuous film. It is known that the coated layer has a porous structure. This feature of the coated layer has significant influence on printing performance parameters. In addition, printing is a dynamic process. Due to difficulty in measuring the dynamic mechanical properties of such a thin and composite layer, dynamic mechanical analysis of the coated layer has not been studied in the paper coating processes. Results based on DMTA studies indicate that three transitions of dynamic viscoelastic properties are observed in the coated layer. The first transition corresponds to polymer T_g and frequency dependent, while second and third transitions show independence on frequency and seem to be related to the packing state of the coated layer. The effect of these transitions related to the printability influenced by latex properties is discussed.

Surface Phenomena and Latexes in Waterborne Coatings and Printing Technologies, Edited by M.K. Sharma, Plenum Press, New York, 1995

INTRODUCTION

There are three basic materials are used in paper coatings and printing processes. These include: (1) pigments, (2) the binder used for the pigment binding and film forming on the substrates, and (3) water which acts as carrier. The pigments for paper coatings used are clay and calcium carbonate. The binders are mainly latex and a small amount of starch or CMC. Typical formulation for paper coating is presented in Table I.

The pigment volume content of the coated layer in paper coating is over critical pigment volume concentration (CPVC) and latex polymer forms a discontinuous film. Therefore, the coated layer forms a porous structure[1,2] . Figure 1 shows a schematic diagram of the coated layer.

Table I. Typical Pigment Coating Formulation for Paper Coating.

MATERIALS	NET PARTS (WT)
Kaolin Clay	70.0
Calcium Carbonate	30.0
SB-Latex	12.0
Starch	3.0

This feature of the coated layer has some significant influence upon the printability[3] of coated paper. The printing parameters influenced include: ink receptivity, surface strength, gravure printability, blistering resistance, and so on through the structure forming process. The structure of the coated layer has been observed and analyzed using an Electron Microscope[6,7], ESCA[8,9], FT-IR[10,11], mercury porosimeter, and several other techniques. The relationships between structure of the coated layer and printability have been discussed. These studies involved commonly used static measurements. No dynamic measurements have been conducted, even the printing is a dynamic process.

Therefore, dynamic investigations on the coated paper are necessary in order to establish a relationship between the structure of the coated paper and printability. There has been no investigations so far to study the dynamic mechanical properties of the coated layer in relation to printability.

Latex, which is quantitatively the component subsequent to pigment in paper coating, is a polymer (carboxylic modified styrene-butadiene copolymer) possessing dynamic viscoelastic properties[12]. It makes sense to consider that the coated layer containing latex polymer has dynamic viscoelastic properties. It is a matter for consideration that the dynamic viscoelastic properties of the coated layer are related to printability. But dynamic viscoelastic analysis has not been applied to practical study of paper coatings, due to difficulty in measuring the viscoelastic properties of such a thin and composite layer. Dynamic mechanical thermal analyzer (DMTA) is a suitable equipment to measure the dynamic mechanical properties of the coated layer (e.g. thin film). It is planned to examine in this paper that how the dynamic mechanical properties of the coated layer are related to the structure depending on latex properties and how this structure affects the printing parameters.

EXPERIMENTAL

MATERIALS

Eight types of latexes were investigated in this study. The properties of these latexes are listed in Table II.

DMTA MEASUREMENTS

Dynamic mechanical analysis of rubbers, polymers and composites over a wide range of temperatures and frequencies provides detailed information about the chemical and physical structure of these materials and their performance characteristics. Dynamic mechanical analysis was done using the bending mode of DYNAMIC MECHANICAL THERMAL ANALYZER (DMTA), made by the Polymer Laboratories. Mechanical head as bending mode of DMTA apparatus is illustrated in Figure 2. Variations of the dynamic storage modulus (E') and the damping factor (tan δ) with temperature and frequency allow characterization of the viscoelastic properties of particular material samples.

SAMPLE PREPARATION FOR DMTA

To measure the pure dynamic mechanical properties of latex polymer films and the pigment coated layer, which contains only clay and latex, these samples were casted on 50 μM polyimide film, not on base paper. Samples of the pigment coating were cured at 150 °C, and 60 seconds after having been

Table II. Properties of Latexes Used in This Study.

LATEX	Particle Size[a] (nm)	T_g[b] (^{0}C)	Gel Content[c] (%)
Latex-STD	140	10	40
Latex-LTG	130	-20	71
Latex-MTG	130	0	70
Latex-HTG	130	20	78
Latex-LGL	150	10	40
Latex-HGL	150	2	90
Latex-SPS	75	10	48
Latex-LPS	150	10	53

[a]Particle Size was Measured by Laser Light Scattering.
[b]T_g was Measured by Differential Scanning Calorimeter (DSC).
[c]Gel content means the weight of insoluble fraction when the film of latex (pH=8) is dissolved in toluene.

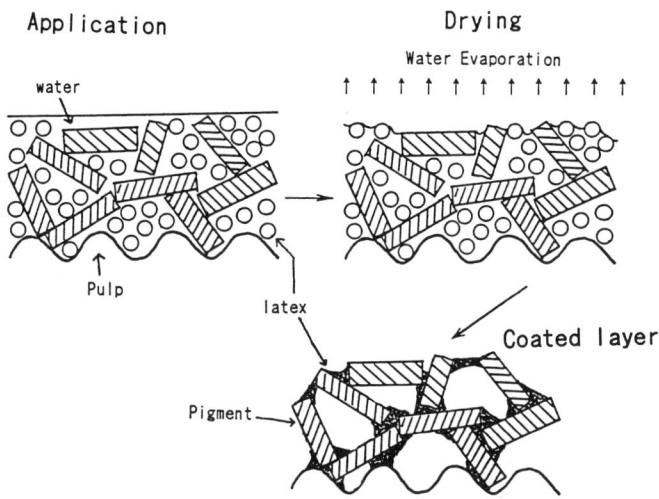

Figure 1. A Schematic Diagram for Forming of Coated Layer.

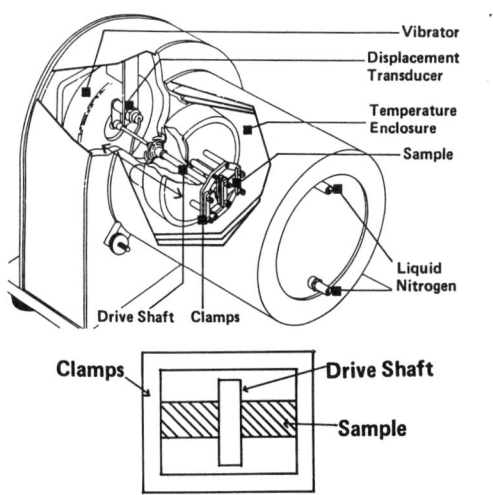

Figure 2. A Schematic Diagram for Mechanical Head of DMTA.

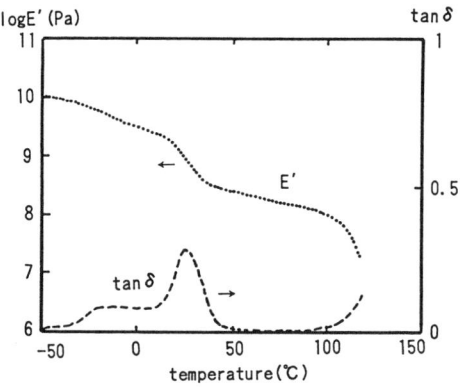

Figure 3. Typical Dynamic Viscoelastic Curve of Latex Polymer.
Ramp Rate: 3°C/Min; Frequency: 3Hz.

kept at room temperature over 24 hours. Film thickness of the pigment coated layer and latex polymers are 200 μM and 350 μM respectively including polyimide film. Polyimide film has no charge in E' and tan δ under experimental conditions from -50 to 150 °C. The effect of the substrate can be negligible.

RESULTS

DYNAMIC MECHANICAL PROPERTIES OF THE PIGMENT COATED LAYER

Typical temperature dependence of dynamic mechanical properties of the latex polymer (Latex-STD) film and coated layer of pigment coating is shown in Figure 3 and Figure 4 respectively.

Temperature dependence of dynamic viscoelastic curve for latex polymer film showed precise definition of a glassy region, transition region, rubbery region and flow region. In the case of coated layer of which the formulation was 10 wt parts latex polymer and 100 wt parts clay (latex/pigment ratio is 20/80), it showed abnormal viscoelastic behavior. Two types of transformations were observed. The former transformation around room temperature was accompanied with E' downward (it is defined as the first transition) and upward (the second transition) transition. The last transformation showed E' downward transition (the third transition) at higher temperatures.

Effect of Latex/Pigment Volume Ratio: Figure 5 illustrates the effect of temperature on dynamic mechanical properties of the coated layer which contains different ratios of latex and clay. Latex-STD was used as a binder for the coated layer. First transition of E' occurred at the same temperature. The second upward transition of E' was gradually decreased when latex/pigment volume ratio was varied, and the upward transition of E' was not observed over 50/50 latex/pigment volume ratio. Dynamic mechanical properties of the coated layer at 65/35 latex/pigment volume ratio was similar to that of the latex polymer.

The third transition of E' was gradually decreased and disappeared around 50/50 of latex/pigment volume ratio. Also the tan δ change was decreased with increasing latex/pigment volume ratio.

Effect of Frequency: Figure 6 illustrates the variations in dynamic mechanical properties for the coated layer as a function of temperature at different frequencies (e.g. 0.3Hz, 3Hz and 30Hz. The ratio of Latex-STD/pigment volume was kept constant (e.g. 20/80). The first downward transition of E' around room temperature shifted to higher temperature. The transition was the same as typical behavior of polymer viscoelastic properties of the coated layer.

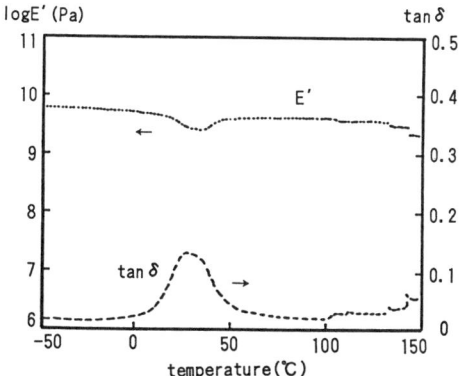

Figure 4. Typical Dynamic Viscoelastic Curve of the Pigmented Coated Layer. Ramp Rate: 3°C/Min; Frequency: 3Hz.

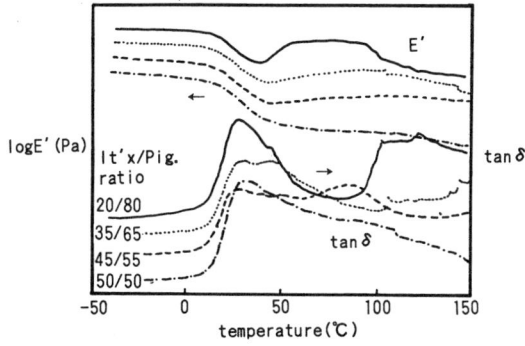

Figure 5. Dynamic Viscoelastic Curves of the Coated Layers with Different Latex/Pigment Ratios. Ramp Rate: 3°C/Min; Frequency: 3Hz.

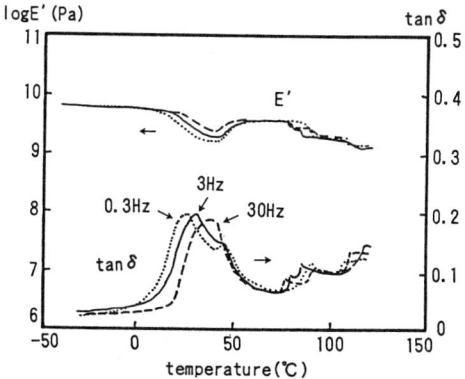

Figure 6. Dynamic Viscoelastic Curves of the Coated Layers at Different Frequencies. Ramp Rate: 3°C/Min.

The second upward transition of E' around room temperature did not shift and was independent on frequency, whereas the third downward transition of E' also seems not to have any shift. In addition, the second and third transitions of E' did not show any frequency dependence.

EFFECT OF POLYMER PROPERTIES ON THE PRINTABILITY USING DYNAMIC MECHANICAL MEASUREMENTS

The effect of polymer properties such as glass transition temperature (T_g), gel fraction and latex particle size on the printability using dynamic mechanical properties is presented in this section.

Effect of Latex Polymer T_g: The dynamic mechanical properties as a function of temperature for the coated layers, which contained latex polymers of different T_g with almost constant gel content, are discussed. Latex-LTG, Latex-MTG and Latex-HTG were used at the constant latex/pigment volume ratio of 20/80. The latex polymer T_g was varied by changing the butadiene/styrene ratio in the latex.

The first E' downward transition and tan δ shifted to higher temperature with increasing polymer T_g. The second upward transition of E' did not shift and was independent of T_g of latex polymer, while the third E' downward transition occurred at the same temperature in spite of latex polymer T_g. The valley around room temperature was decreased with increasing T_g of latex polymer.

Effect of Gel Content: The variations in dynamic mechanical properties of the coated layer as a function of temperature with different gel fraction of latex polymers are presented in Figures 8 and 9. The gel content of latex polymer was controlled by chain transfer reagent. Latex-LGL and Latex-HGL were used and the latex/pigment volume ratio was 20/80.

The first and the second transitions of E' remained essentially unchange, whereas the third transition of E' was strongly influenced by the gel content. The third transition of the coated layer of Latex-HGL occurred at a higher temperature than that of the lower gel (Latex-LGL) coated layer. From viscoelastic curve, it is clear that the transition at latex polymer melt-flow region is influenced by the gel content.

Effect of Latex Particle Size: Figure 10 illustrates the dynamic mechanical properties for the coated layer versus temperature for latexes with different particle sizes. Latex-SPS and Latex-LPS were used with 20/80 latex/pigment volume ratio. The latex particle size was controlled by emulsifier in the latex formulation. Large particle size latex provided a large and deep valley in the region from the first to the second transition.

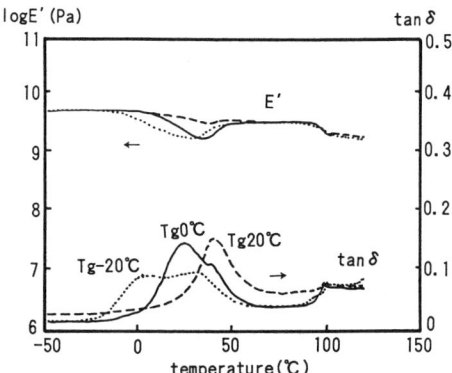

Figure 7. Dynamic Viscoelastic Curves of the Coated
Layers with Different T_g of Latexes.
Ramp Rate: 3°C/Min; Frequency: 3Hz.

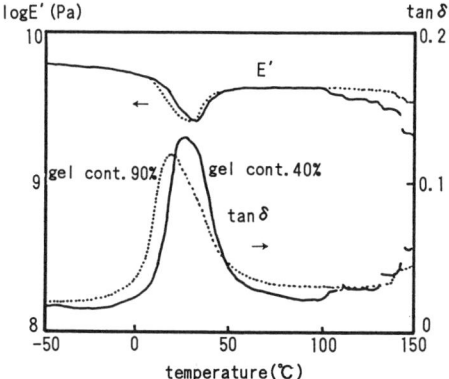

Figure 8. Dynamic Viscoelastic Curves of the Coated
Layers at Different Gel Contents of Latex.
Ramp Rate: 3°C/Min.; Frequency: 3Hz.

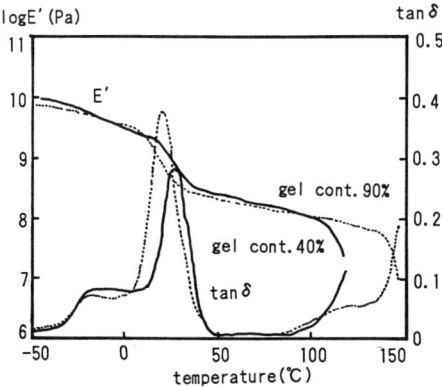

Figure 9. Dynamic Viscoelastic Curves of the Latex
Polymer at Different Gel Contents.
Ramp Rate: 3°C/Min.; Frequency: 3Hz.

DISCUSSIONS

Three different transitions of E' observed during dynamic mechanical properties of the coated layer versus temperature are discussed as follows:

FIRST TRANSITION OF DYNAMIC VISCOELASTIC CURVE

The first downward transition of E' appeared at the same temperature as T_g of latex polymer (Figure 7). Furthermore, it has a frequency dependency on dynamic mechanical measurements as shown in Figure 6. Therefore, the first downward transition of E' is closely related to latex polymer which had frequency dependence and it corresponds to the T_g of latex polymer.

SECOND TRANSITION OF DYNAMIC VISCOELASTIC CURVE

The second transition of dynamic viscoelastic curve showed upward transition of E'. If one relates this transition to latex polymer, this can be an abnormal behavior for latex of styrene-butadiene copolymer, which is not a crystalline polymer.

The second transition of E' disappeared with increasing latex/pigment volume ratio (Figure 5). It was independent of frequency of dynamic viscoelastic measurements as shown in Figure 6. These results suggest that the second transition of E' is not related to latex polymer.

To confirm the occurrence of second transition, experiments were conducted and repeated for temperature scan from -50 to 100 °C in Latex-STD during which melting occurred at 100 °C. Figure 11 shows that the second transition of E' disappeared in the second scan. This viscoelastic curve exhibits the change in the coated structure. It is concluded sense that the packing state of the coated layer seems to close by the orientation of pigment. The possible explanation for various phenomena occurring in the process are schematically presented in Figure 12.

THIRD TRANSITION OF DYNAMIC VISCOELASTIC CURVE

Figures 8 and 9 illustrate that the third transition of the coated layer occurs at a temperature just lower than that of the polymer melt-flow region. Also the third transition is independent of the frequency (Figure 6), like the second transition. The third transition is mainly affected by latex polymer gel content (e.g. gel fraction) which varies polymer melt-flow region behavior. It indicates that this phenomenon occurs by the relaxation of closed packing state of pigment and the deformation of the coated layer near polymer melt-flow region.

PRINTABILITY

It is considered that the frequency independence of the second and third transitions indicates the change of the coated layer structure and is not directly related to the polymer viscoelasticity. However, it is found that these transitions can be affected by latex properties and related to printability of the coated layer as follow:

Table III shows the effect of latex polymer properties on printing performance parameters (e.g. surface strength, ink receptivity, gravure printability, blistering resistance etc.). An attempt is made to establish a correlation among latex polymer T_g, particle size and gel fraction with printability.

Table III. Effect of Latex Polymer Properties on the Printing Performance Parameters

Factors	Surface Strength	Ink Receptivity	Gravure Printability	Blistering Resistance
T_g ⇑	-	...	- - -	...
Gel Fraction	+++	...	-	- - -
Particle Size	-	+++	+	+

+++ Significant Positive Effect
+   Positive Effect
... Unclear
-   Negative Effect
--- Significant Negative Effect

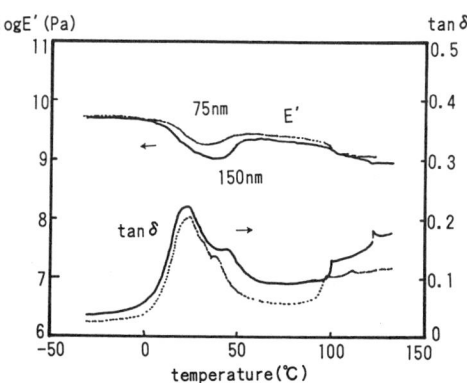

Figure 10. Dynamic Viscoelastic Curves of the Coated Layers with Different Particle Size of Latexes. Ramp Rate: 3°C/Min; Frequency: 3Hz.

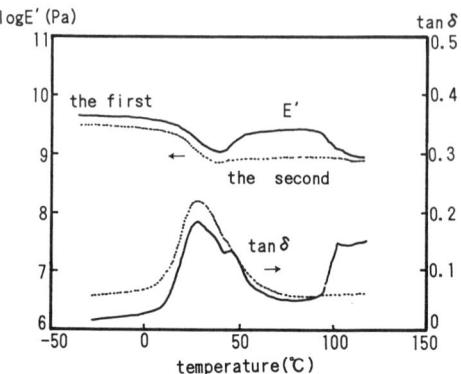

Figure 11. Dynamic Viscoelastic Curves of the Coated Layers at Temperature Scan of Each Cycle. Ramp Rate: 3°C/Min; Frequency: 3Hz.

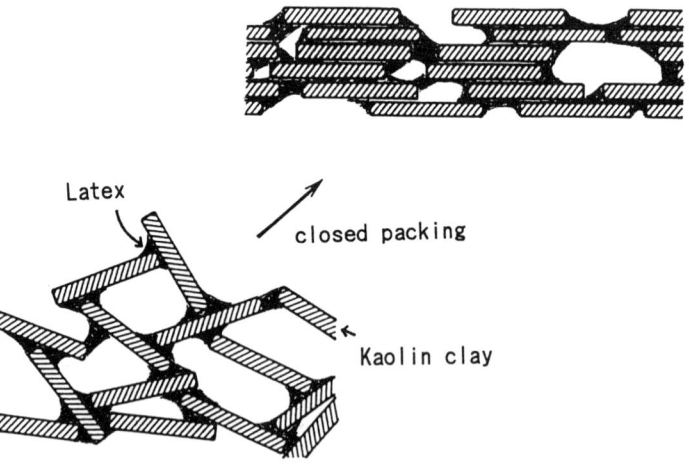

Figure 12. A Schematic Diagram for Closed Packing of the Coated Layer.

GRAVURE PRINTABILITY

Gravure printing is an intaglio printing process. Figure 13 illustrates gravure printing process and intaglio for gravure printing. Light and shade in gravure printing was controlled by depth and area of intaglio cells.

It is important in gravure printing that the surface of the coated paper contacts closely with the coated layer due to good dynamic smoothness. The good dynamic smoothness can be obtained by the compressibility like a cushion of the coated layer. This performance is improved by lower T_g and large particle size of latex.

Area of E' valley from the first transition to the second transition seems to be related to the compressibility of the coated layer. Lower T_g and larger particle size, which improve the gravure printability, provides larger E' valley. It is believed that the E' valleys related to the compressibility of the coated layer.

BLISTERING RESISTANCE

Blistering resistance is an important performance for web-offset printing. In web-offset printing, heat set inks are used. When the ink dries under high temperature over 100 °C, the moisture in paper evaporates similar to an explosion and blistering occurs in the case of not enough permeation of vapor through the coated layer. Figure 14 shows the phenomenon of blistering. To restrain blistering it is necessary to promote permeation of aqueous vapor in the coated layer.

It is known that the latex polymer of lower gel content can provide a better blistering resistance than that of higher gel content in the paper coating. Figure 8 shows that the deformation of the coated layer which contained lower gel content of latex polymer, occurs at a lower temperature than that of higher gel content of latex polymer. The third transition of E' suggests deformation behavior in the region of high temperature, which is related to the blistering resistance.

CONCLUSIONS

Dynamic mechanical properties of the coated layer in paper coating can be measured by DMTA. DMTA provides useful information to characterize the dynamic behavior of the coated layer, which is related to the performance of coated paper.

There are three transitions of E' observed in the measurements of the dynamic mechanical properties for the coated layer. These transitions are summarized as follows: (1) The first downward transition of E' corresponds to latex polymer T_g;

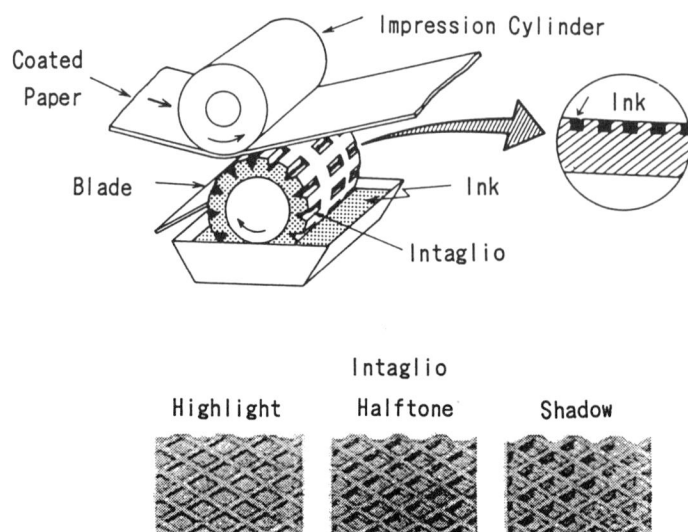

Figure 13. A Schematic Diagram for Gravure Printing Process and Intaglio.

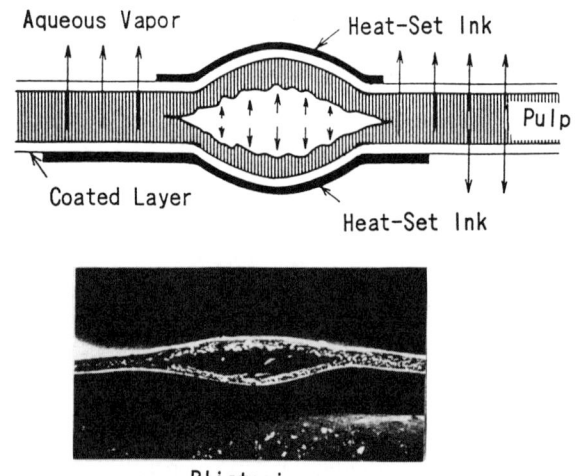

Figure 14. A Schematic Diagram for Blistering Phenomena.

(2) The second upward transition of E' is observed just after the first E' transition and is a result of change in the packing state of the coated layer; and

(3) The third transition of E' occurs at a higher temperature near the melt-flow of latex polymer. This is related to the deformation of the packing structure of the coated layer.

These dynamic mechanical properties of the coated layer are considerably related to the printability of coated paper.

REFERENCES

1. Lepoutre, P. and Allince, B., J. Appl. Polymer Sci., **26**, 791-798, (1981).

2. Lepoutre, P., Progress in Organic Coatings, **17**, 89-109, (1989).

3. Aspler, J.S. and Lepoutre, P., Symposium on Paper coating Fundamentals, 77-98, (1991).

4. Hagymassy, J., Lee, D.I., Schemitt, Givens, S.P. and Hayness Jr., L.U., PAPPI J. **61(1)**, 59-62, (1978).

5. Yamawaki, K., Sasagawa, Y. and Tsuji, A., 1991 TAPPI Coating Conference Proceedings, 199-206, (1991).

6. Matsubayashi, H., Takagishi, Y., Miyamoto, K. and Kataoka, Y., 1990 TAPPI Coating Conference Proceedings, 419-430 (1990).

7. Whalen-Shaw, M. and Eby, T., 1991 TAPPI Coating Conference Proceedings, 401-410, (1991).

8. Engstrom, G., Norrdahl, P. and Strom, G., 1987 TAPPI Coating Conference Proceedings, 35-44, (1987).

9. Fujiwara, H. and Kline, J.E., 1986 TAPPI International Process and Material Quality Evaluation Conference Preprint, 157-164, (1986).

10. Reif, L., Seelemann, R. and Wallpot, G., 14th PTS Coating Symposium, Munchen, 80-83, (1989).

11. Fujiwara, H. and Kline, J.E., 1987 TAPPI Coating Conference Proceedings, 29-34, (1987).

12. Ferry, J.D., Viscoelastic Properties of Polymers, Willy, New York/London, 3rd Edition, (1961).

SOLIDS DETERMINATION IN LATEX AND

LATEX-BASED COATINGS

Bradley H. Larson

Arizona Instrument Corporation
Phoenix, Arizona 85040-1941

ABSTRACT

Solids determination is crucial to the correct formulation of latex-based coatings. Generally, too high a level of solids can result in degradation of physical properties, while a solids content which is too low can contribute to problems in both surface and pigment binding. Although standard oven procedure has been a commonly accepted method of solids determination, test times are generally very lengthy and therefore not suitable for spot-checking formulations during processing. An alternative method is now available providing quick, accurate solids determinations in both latex and latex-based coatings. Using an electronic force balance to continuously monitor weight loss during heating, a microprocessor interprets the information and compares the sample's weight loss to a standard drying curve. The final solids value is then extrapolated from the curve and results are available within minutes. This method gives results in direct correlation to the oven method, but in a fraction of the time. It is suitable for the laboratory as well as the production environment.

INTRODUCTION

Quick, accurate solids analysis is important in the production of latex and latex-based coatings for several reasons. In the formulation of latex polymers, it is important to drive the monomer conversion to the highest level possible. The ability to quickly reduce residual monomer

Surface Phenomena and Latexes in Waterborne Coatings and Printing Technologies, Edited by M.K. Sharma, Plenum Press, New York, 1995

107

concentration to less than 0.1% is of considerable commercial importance. It results in reduced total process time while improving overall product quality.[1] Therefore, solids analysis can be used as a tool in timing the duration of process. In the processing of latex-based coatings problems with surface and pigment binding, as well as degradation of physical properties, can occur as a result of improper solids levels. While the standard oven test[2] has been used in the past as a reliable test for solids, an instrument is now available which gives accurate results in a fraction of the time. That instrument is the Computrac MAX-50 and it is the subject of this paper.

EXPERIMENTAL

APPARATUS AND METHOD

The Computrac MAX-50 is a microprocessor-controlled loss-on-drying (L.O.D.) system designed to measure moisture and/or solids levels by continuously monitoring loss in weight and comparing that loss to a standard drying curve. Sample is added to an aluminum sample pan within a test chamber. The aluminum pan sits on a pan support coupled to an electronic force balance. The force balance registers the initial weight of the sample prior to testing and, upon heating, relays the constant decrease in sample weight to the microprocessor for calculation. Heating is done by a 700 watt nichrome element which is affixed to the underside of the test chamber's lid. The temperature of the test chamber is maintained within 2°C of the programmed temperature throughout the test by use of an RTD (resistance temperature device) also mounted on the underside of the lid.

As the microprocessor continuously monitors the sample's weight loss during testing, the decrease in weight is compared to the initial sample weight and the calculated solids concentration appears on the display. Simultaneously, the microprocessor starts a prediction of the final solids concentration based upon the sample's rate of weight loss compared to the exponential portion of a standard drying curve. The test will finalize when the predicted solids concentration agrees within a certain percentage of the actual concentration appearing on the display. The percent agreement between the predicted and the actual solids concentration will vary depending on the system's programming and the moisture level of the sample.[3]

A standard drying curve consists of three portions (Figure 1). The first portion (A-B) represents the sample's weight loss as it heats from ambient to testing temperature. Section B-C is linear, representing volatile loss from the sample including free as well as bound moisture. The final portion of the curve (C-D) is exponential in nature and represents the final evolution of volatiles. The point where no additional weight loss occurs represents the complete evolution of volatiles from the original sample (E). The time for a sample to reach zero additional weight loss can take from one to 24 hours depending on the sample, sample preparation and testing temperature. The MAX-50 calculates the sample's solids concentration through mathematical extrapolation from the exponential (C-D) portion of the curve.

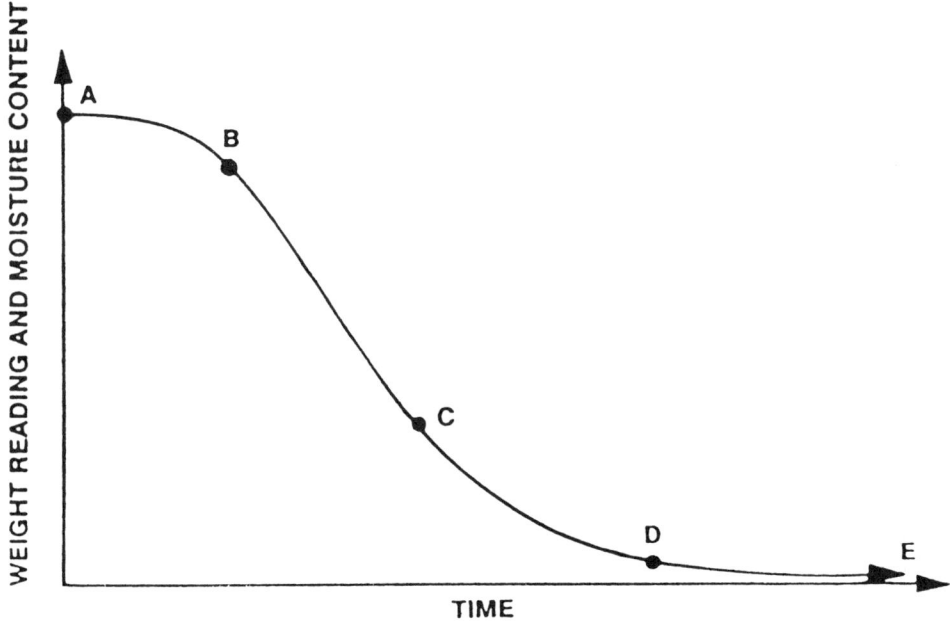

Figure 1. Drying Curve

The MAX-50 does an extrapolation through a process of weight averaging. Individual weight readings are taken every 0.75 seconds and once eight readings are received they are averaged and stored as one set. Once four of these sets are received they are then averaged and used to predict a final endpoint.

All weight monitoring is done by the use of an electronic force balance. The balance has a capacity of 40,000 electrical "counts". Each count has a percentage value of approximately .0025%. Variance in prediction is determined by the ending criteria (EC) which is the value given to the difference between the predicted and actual solids concentration. Four different EC's are available: 500, 255, 97 and 06.

Table I. MAX-50 solids values for Acrylic Latex

SAMPLE SIZE, %	% SOLIDS	TEST TIME (MINUTES)	TEST TEMP. (°C)
25	53.64	5.18	180
25	53.69	5.15	180
26	53.65	4.52	180
25	53.56	4.54	180
25	53.76	4.31	180
24	53.59	4.10	180

Mean = 53.65%
SD =   .065%

TMAX software was used during one of the tests. TMAX is a data collection program designed to collect data from the Computrac MAX-50 and graph this data in real time using a personal computer (Figure 2). The results are given as percent moisture and the inverse of these values can be taken to represent percent solids (Table I).

PROCEDURE

In order to establish a known solids value for comparison, a standard oven test was performed. Ten samples were prepared using approximately 1-2 grams of an acrylic latex polymer. Each sample was placed in an aluminum sample dish and then put in a forced air oven heated to 105°C. The samples were removed after three hours and then weighed. A total solids value was then calculated for each sample and the mean and standard deviation were figured.

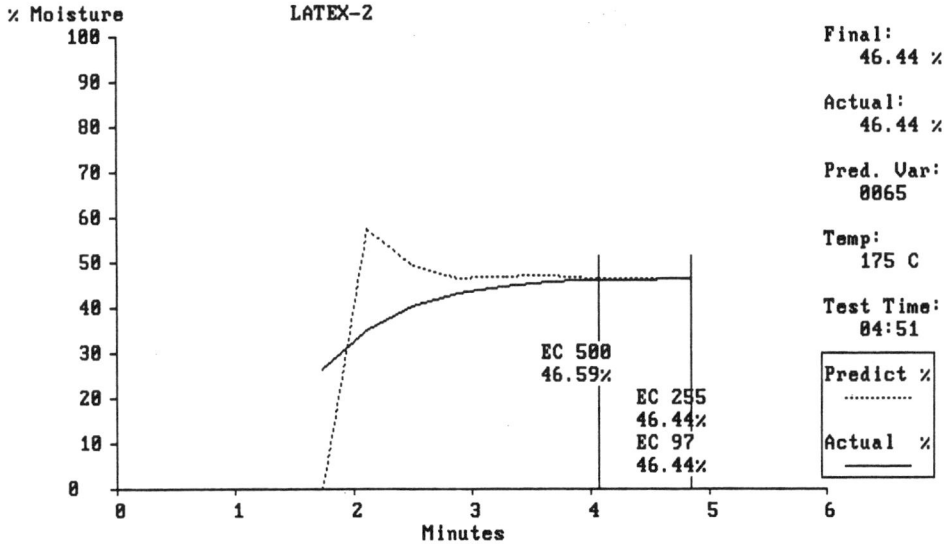

Figure 2. TMAX Graph

A total of six tests were run using the Computrac MAX-50 Moisture Analyzer. Approximately 2.5 grams of sample were used per test. The sample was placed on a glass fiber sample pad in order to lessen the chance of bubbling during heating. All testing was done at a set temperature of 180°C with a selected ending criteria of 97. In order to reduce the level of initial volatile loss, the chamber temperature was allowed to cool to 38-40°C between sample loadings. The acrylic latex polymer used in the study was periodically shaken in order to keep the sample in suspension.

RESULTS AND DISCUSSION

The solids values obtained using the MAX-50 ranged from
53.56% to 53.76%, with a mean statistical value of 53.65% and
a standard deviation of .065% (Table II). Test times were 4-5
minutes.

Solids values using the oven method ranged from 53.79% to
53.95%, exhibiting a statistical mean of 53.89% and a standard
deviation of .059% (Table III). Test time for the oven method
was approximately three hours.

Table II. Oven solids results for Acrylic Latex

PAN WEIGHT	PAN+SAMPLE WEIGHT	SAMPLE WEIGHT	PAN+DRY SAMPLE WEIGHT	% SOLID
1.1371	2.6263	1.4865	1.9374	53.84
1.1191	2.3291	1.2100	1.7718	53.94
1.1456	3.1304	1.9848	2.2150	53.88
1.`322	2.2257	1.0935	1.7204	53.79
1.1312	2.7099	1.5787	1.9827	53.94
1.1374	2.2090	1.0716	1.7155	53.95

Mean = 53.89%; SD = 0.059%

Table III. TMAX Values

File Name: LATEX-2.MAX   Page: 1

```
==================================================================
0000 PRED. VAR   00.00% PREDICT  176 DEG C  26.72% ACTUAL  01:44 TEST TIME
9999 PRED. VAR   57.64% PREDICT  181 DEG C  35.17% ACTUAL  02:07 TEST TIME
9999 PRED. VAR   49.51% PREDICT  180 DEG C  40.49% ACTUAL  02:30 TEST TIME
5557 PRED. VAR   46.63% PREDICT  175 DEG C  43.34% ACTUAL  02:54 TEST TIME
1153 PRED. VAR   46.85% PREDICT  173 DEG C  44.91% ACTUAL  03:17 TEST TIME
0377 PRED. VAR   47.21% PREDICT  175 DEG C  45.84% ACTUAL  03:41 TEST TIME
0356 PRED. VAR   46.59% PREDICT  175 DEG C  46.25% ACTUAL  04:04 TEST TIME
0348 PRED. VAR   46.46% PREDICT  175 DEG C  46.39% ACTUAL  04:28 TEST TIME
0065 PRED. VAR   46.44% PREDICT  175 DEG C  46.44% ACTUAL  04:51 TEST TIME
```

CONCLUSION

The Computrac MAX-50 is a method which can provide quick, accurate solids determinations in latex and latex-based coatings. The MAX-50 correlates well with the standard oven test, and because of its short test times (generally under six minutes), it can be used for spot-checking in areas where the oven test is simply too time consuming to be efficient.

REFERENCES

1. Kamath, V.R. and Sargent, J.D. Jr., J. Coatings Technology, **59**, 53, (1987).

2. ASTM D2369-87·

3. Arizona Instrument Corp., MAX-50 Operations Manual.

IMPROVEMENT OF WET ADHESION OF ORGANIC COATINGS

BY THIN ADHESION LAYER

W. Funke

2nd Institute for Technical Chemistry
University of Stuttgart and Forschungsinstitut fur
Pigment und Lacke e.V.
D-7000 Stuttgart 80
Germany

ABSTRACT

An attempt was made to discuss several mechanisms of wet adhesion to improve the performance properties of organic coatings against corrosive and destructive environment. The effect of cooperative bonding to improve immobilization during film formation by crosslinking between binder and substrate molecules was presented. It was demonstrated that the wet adhesion increased by increasing glass transition temperature (T_g) of the polymer. A proper confirmation of binder molecules in very thin adhesion layers increases the bonding efficiency resulting the improvement of wet adhesion of the organic coatings on the substrates. The combination of crosslinkable macromolecules used as a binder with barrier properties of the thin adhesion layer of coating can provide an interesting alternative for the common electrochemical protection for heavy duty applications.

Surface Phenomena and Latexes in Waterborne Coatings and Printing Technologies, Edited by M.K. Sharma, Plenum Press, New York, 1995

115

INTRODUCTION

One of the most important functions of organic coatings is protection of substrates against corrosive and destructive influences of the environment.

Two mechanisms may serve this purpose: (1) the electrochemical mechanism e.g. by anticorrosive pigments or inhibitors, and (2) the barrier mechanism which prevents or retards the permeation of aggressive substances, such as water, oxygen or ions, through the coating film to the substrate. For the barrier mechanism to operate efficiently, it is important that organic coatings resist the delaminating action of water. The adhesion of organic coatings on exposure to water - wet adhesion - may be improved by barrier pigments, by promoting the cooperative bonding to the substrate and by improving the conformation of the macromolecules from the binder, which are in direct contact with or are adsorbed at the substrate surface.

The improvement of the protective quality by barrier pigments is achieved by a prolongation of the diffusion pathways. The effect of such pigments has been described earlier[1].

An attempt is made to discuss two additional mechanisms, which are directly involved in wet adhesion.

THEORETICAL CONSIDERATIONS

The interactions and their role in wet adhesion of the organic coatings are presented briefly in this section.

COOPERATIVE BONDING

Adhesion of Organic coatings to polar substrates, such as metals, depends on polar interactions. These interactions may be easily disturbed by water molecules. As such polar interactions are needed for adhesion, it was considered how to make them resistant to the delaminating action of water.

Polar interactions are relatively weak bonds as compared with, e.g. covalent bonds. Such weak bonds are normally subjected to an equilibrium between the bonded and non-bonded state. This equilibrium can be shifted to the bonded state and stabilized by making the chain segments rigid, to which the bonding groups of the binder molecules are attached.

However, in order to enable the adsorption process during film application and formation, the immobilization should occur during film formation by crosslinking reaction. Crosslinking immobilizes the interfacial bonds and thus enforces cooperation between them as shown in Figure 1.

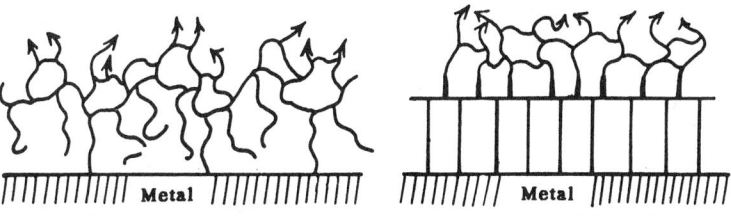

Arrows indicate the continuation of chain segments into the bulk
of the coating

Non-cooperating adhesion bonds.
Dynamic equilibrium between
bonding and non-bonding chain
ends.
Mobile chain segments at the
interface

Cooperating adhesion bonds.

Immobile chain segments at
the interface.

Figure 1. Non- Cooperative and Cooperative Adhesion Bonds
at the Coating/Support Interface.

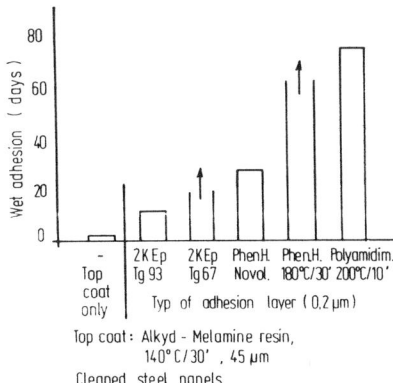

Figure 2. Influence of Thin Adhesion Layers (0.2 μm) on
Wet Adhesion of an Alkyd-Melamine Resin Film.

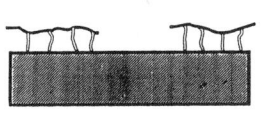

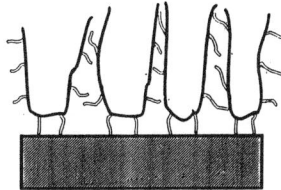

uncomplete coverage
(very low concentration)

complete coverage, but
less ordered than on water
(low concentration)

Overcoverage
(high concentration
as in practical coatings)

Figure 3. Concentration Effects on Molecular Order and
Conformation of Polar Macromolecules at
Polar Substrates.

RESULTS AND DISCUSSION

The immobility of polymer chain segments is reflected by the glass transition temperature. One should expect an increase of wet adhesion with increasing T_g. As is shown in Table I, wet adhesion actually agrees mostly with glass transition temperatures[2]. Exceptions can be explained by the plasticizing action of water.

It is shown experimentally as illustrated in Figure 2 that wet adhesion of an alkyd/melamine resin coating on various rigid adhesion layers of only 0.2 um film thickness applied to steel panels is substantially improved[2,3].

CHAIN CONFIRMATION AT SUBSTRATE SURFACES

It is known that the normal paints are highly concentrated solutions of binders. As is known from adsorption studies, e.g. at pigment surfaces, such high concentrations do not allow a regular formation of adsorbed layers of macromolecules, because they compete with each other for the bonding sites at the substrate surface. In order to permit an optimal yield of adsorption sites per macromolecule, the binder concentration has to be reduced strongly. One may expect a certain concentration, at which the coverage of the substrate is complete though less ordered than a monolayer on a mobile aqueous surface (Figure 3).

It could be demonstrated that such a layer can be prepared with polyacrylic acid and the surprising result was, that within a small concentration range the wet adhesion of a normal coating applied on such very thin adhesion layers increases by several orders of magnitude in sizes. It has been shown[4] that the wet adhesion decreases continuously in the high concentration region as presented in Figure 4. Similar results were obtained with polymethacrylic acids as illustrated in Figure 5. In Figure 6 it is shown that wet adhesion of epoxy and polyurethane coatings increases similarly if thin polyacrylic acid adhesion layers have been applied. Similar results were reported in the literature[5].

By plotting the weight change (ΔG) of the steel panels, as observed by the pretreatment with polymethacrylic acid, against the concentration of the polyacid solutions at variable numbers of immersions (Figure 7) the crossing point of the curves appears at a concentration of 1.5% by mass, which means that this is the optimal concentration for the monolayer formation in which the macromolecules at the interface have a maximal number of adsorptive groups placed at the substrate[5]. A comparison with Figure 5 shows, that this concentration is within the range, in which maximal wet adhesion was observed.

It should be emphasized, however, that the stability of polyacid adhesion layers alone against water is limited. Overcoating with aqueous top layers does not usually improve wet adhesion. Adhesion layers and top coats have to be adjusted to each other.

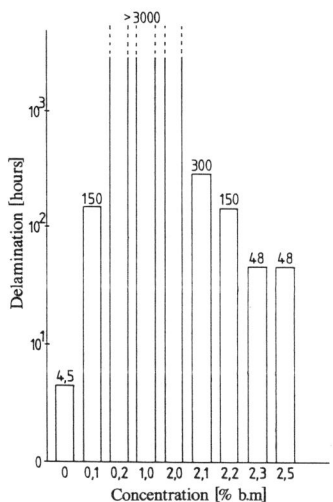

Figure 4. Wet Adhesion (Delamination Time) and Concentration of Aqueous Polyacrylic acid (Molar Mass 70,000) Solutions.
Top Layer: Alkyd-Melamine Resin - 40 μm.

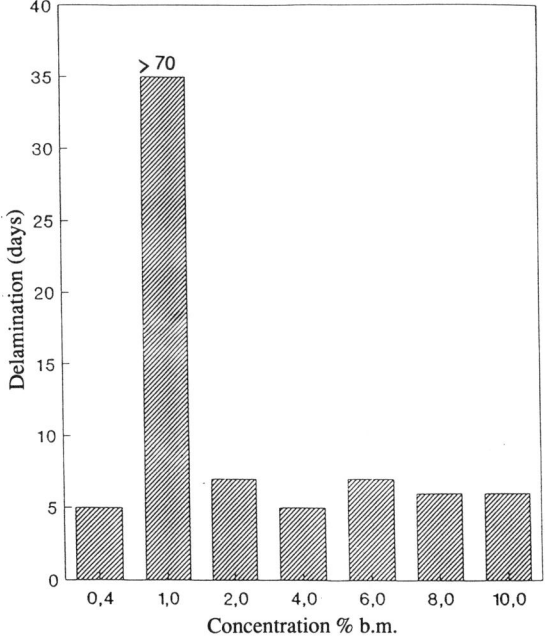

Figure 5. Wet Adhesion (Delamination Time) and Concentration of Aqueous Polymethacrylic acid (Molar Mass 70,000) Solutions.
Top Layer: Alkyd-Melamine Resin - 40 μm.

Table I. Wet Adhesion and Glass Transition Temperature (T_g) of Organic Coatings in the Dry State.

Coating No.	Wet adhesion (hrs.)	T_g (°C)
I	1880	88
II	1880	83
III	1008	82
IV	120	30/83
V	18	44
VI	8	82
VII	1,5	20
VIII	1,0	27

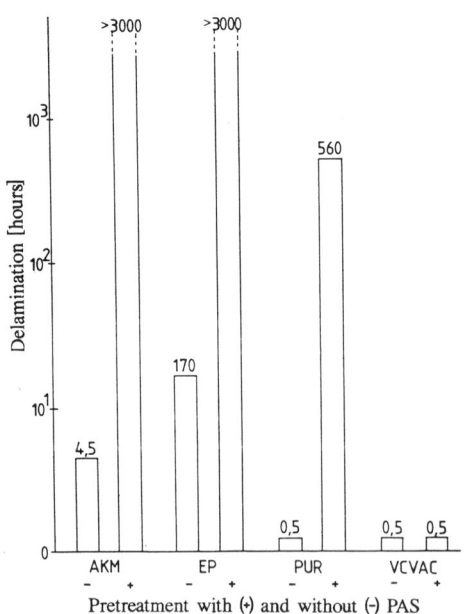

Figure 6. Wet Adhesion of Different Top Layers (40 µm) on a Polyacrylic Acid Adhesion Layer.

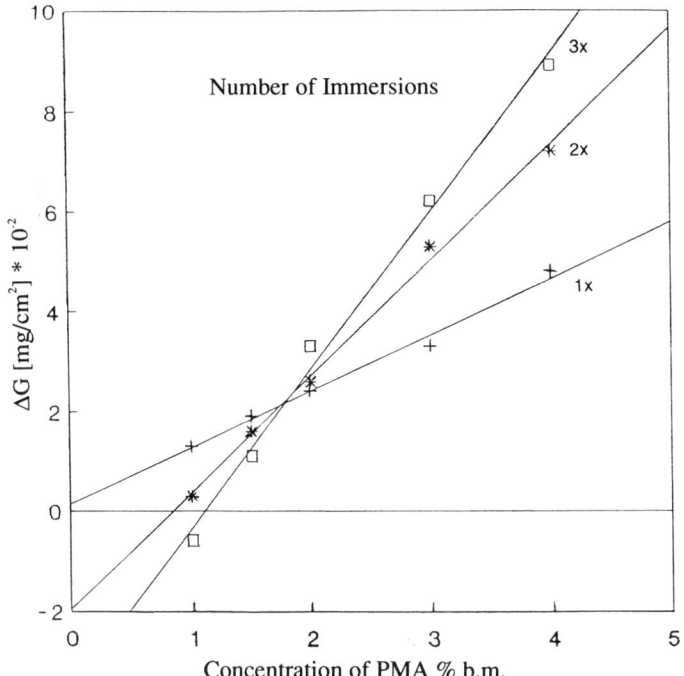

Figure 7. Mass Change (ΔG) of Steel Panels Pretreated
with Aqueous Polymethacrylic Acid Solutions
of Variable Concentration and Number of
Immersions (1 Min Each).

CONCLUSIONS

It was demonstrated that the cooperative bonding improves
wet adhesion and delays delamination by water. A suitable
conformation of macromolecules in very thin adhesion layers
(about 10.0 nM) increases the bonding efficiency and wet
adhesion.

Thin adhesion layer of special crosslinkable
macromolecules combined with barrier properties of the organic
coating following the adhesion layer is an interesting
alternative of the common electrochemical protection,
especially in case of heavy duty conditions.

REFERENCES

1. Funke, W., J. Coatings Technology, **55**, 31 (1983).

2. Funke, W., Symposium on Advances in Corrosion Protection by
   Organic Coatings, and The Electrochem. Soc. Pro. Vol. 89,
   13, Edited by D. Scantlebury and M. Kendig, Cambridge, UK,
   (1989).

3. Arslanov, V. V. and Funke, W., Progress in Organic Coatings, **15**, 365, (1988).

4. Funke, W. and Yamabe, H., J. Japan Soc. of Color Material, **64**, 630, (1991).

5. Gao, Z., Yamabe, H., Marold, M. and Funke, W., H. Forbe and Lack, **98**, 917, (1992).

DYNAMIC WETTING OF WATER-BASED INKS IN FLEXOGRAPHIC AND GRAVURE PRINTING

F.J. Micale, S. Sa-Nguandekul,
J. Lavelle and D. Henderson

Sinclair Lab.
Lehigh University
7 Asa Drive
Bethlehem, PA 18015

ABSTRACT

The theory of wetting is reviewed with respect to ink transfer which is based upon measured dynamic surface tension and calculated dynamic spreading coefficient. Laboratory gravure ink transfer results are presented for model water based inks with and without isopropanol as the cosolvent on untreated and corona treated polyethylene film. A mechanism of surface tension driven convection is proposed which is consistent with experimental results. The conclusion, which is based upon the proposed mechanism, is that uniform coverage of a water based ink on a nonpermeable substrate is facilitated by the presence of a high vapor pressure low surface tension cosolvent such as isopropanol. When no cosolvent is present, de-wetting and degree of ink mottling appears to be controlled by dynamics longer than one second.

INTRODUCTION

This paper is the third in a series on the role of wetting in printing where the first paper investigated the role of wetting in lithographic printing[1], and the second paper reviewed some of the implications of dynamic surface tension and presented results of laboratory flexographic ink transfer as a function of equilibrium surface tension and

Surface Phenomena and Latexes in Waterborne Coatings and Printing Technologies, Edited by M.K. Sharma, Plenum Press, New York, 1995

wetting[2]. This paper will present results based on laboratory gravure printing. It will review wetting theory as determined from contact angle measurements and will present calculations of solid/liquid interfacial tension measurements. The dynamics of surface tension and wetting are additional concepts which must be understood in order to define a mechanism which is capable of relating wetting results to ink transfer. A correlation between ink/substrate wetting characteristics and printability depends on the reliability of the laboratory ink transfer technique which is used to generate test prints. A further refinement of the modified Prufbau Printability Tester[3] will be presented as an important element of the experimental results which are intended to give reliability to the print film results.

The primary printing problem addressed in this paper is the wetting of model water based gravure printing inks, which are formulated with and without isopropanol as the cosolvent, on nonpermeable substrates such as film-base materials. Wetting is much more critical on nonpermeable substrates where there is essentially no immobilization of the ink film after ink transfer. Water, furthermore, is a highly polar solvent with a high surface tension which exhibits pronounced dynamic surface tension properties compared to organic solvents. Although water based inks can be formulated to yield equilibrium surface tension and wetting values similar to solvent inks, uniform ink coverage is usually more difficult to achieve for the water based inks. The reason is that ink transfer in the nip to final ink curing occurs in fractional seconds where the non-equilibrium surface tension of the water based inks can undergo large changes in surface tension values. Experimental approaches and concepts designed to understand the nature of this problem will be proposed.

THEORETICAL CONSIDERATIONS

THEORY OF WETTING[4]

A fundamental concept of wetting is the spreading coefficient of a liquid on a solid surface, $S_{1/s}$, defined according to the following equation:

$$S_{1/s} = \gamma_s - \gamma_{sl} - \gamma_1 \qquad (1)$$

where γ_s = the surface tension or energy of the solid,

and γ_{sl} = the solid/liquid interfacial tension or energy,

γ_1 = the surface tension or energy of the liquid.

If all the terms in Equation 1 are taken as energy terms, then spreading will result in a lower energy for the system if γ_s is greater than the sum of γ_{sl} and γ_1. A positive value for the spreading coefficient, therefore, indicates spontaneous spreading and a negative value indicates non-spreading. The magnitude of the positive or negative value, furthermore, is a measure of the driving force for the liquid to spread.

Equation 1 presents a problem because the only term which can be measured readily is the surface tension of the liquid. A basic equation of wetting is the Young Dupre Equation which defines the balance of surface tension forces and contact angle, θ, for a nonspreading liquid drop at equilibrium with a surface:

$$\gamma_s = \gamma_{sl} + \gamma_1 \cos \theta \qquad (2)$$

where θ is the angle between the liquid and the solid at the gas/solid/liquid interface. Equation 2 presents a problem in the sense that the only terms which can be measured readily are γ_1 and θ. Combining Equations 1 and 2 results in Equation 3:

$$S_{1/s} = \gamma_1 (\cos \theta - 1) \qquad (3)$$

where all the terms for evaluating the spreading coefficient can be measured experimentally. The spreading coefficient has the significance, in general, of predicting the degree of ink-substrate interaction or tendency for wetting where a high positive spreading coefficient can lead to ink penetration and feathering, and a high negative spreading coefficient can lead to poor ink transfer and nonuniform coverage.

The geometric mean theory[5] defines the interfacial tension, γ_{sl}, between a solid and a liquid, or between two immiscible liquids, according to the following equation:

$$\gamma_{sl} = \gamma_s + \gamma_1 - 2 (\gamma_s^d \gamma_1^d)^{1/2} - 2 (\gamma_s^p \gamma_1^p)^{1/2} \qquad (4)$$

where

γ_s = the surface tension of the solid,

γ_1 = the surface tension of the liquid,

γ_s^d = the dispersion component of surface tension of the solid,

γ_1^d = the dispersion component of surface tension of the liquid,

γ_s^p = the polar component of surface tension of the solid,

γ_1^p = the polar component of surface tension of the liquid.

The dispersion components of surface tension are non-specific in the sense that the dispersion component of surface tension of one material will always undergo an attractive interaction with the dispersion component of surface tension of another material. The geometric mean theory predicts that the interfacial tension between two materials will be reduced by 2 X the square root of the multiple of the dispersion components of surface tension for the two materials, i.e. the third term on the RHS of Equation 4. The polar components of surface tension, however, are by definition specific in nature. The types of interactions which may result from polar components include acid-base, hydrogen bonding, and polar interactions. Since polar interactions are specific in nature, an attractive interaction may or may not occur between the polar components of the solid and the liquid. For example, a strongly acidic polar component would be expected to undergo total attractive interaction with a strongly basic polar component, and zero attractive interaction with a strongly acidic polar component. All levels of intermediate degrees of interaction, of course, are possible. The implication is that the coefficient 2 in the last term on the right hand side (RHS) of Equation 4 can in fact vary from 0 to 2.

DYNAMIC SURFACE TENSION

There is no question that dynamic surface tension plays an important role in ink transfer and printability for all printing processes. An ink is exposed to a surface and undergoes splitting in the nip in fractional seconds at which point an ink/substrate and ink/air interface are generated. An important consideration for water based inks is the concept that the surface tension for any surfactant solution at zero time, i.e. at the instant of generation of a new surface, will be equal to the surface tension of pure water. This concept is based upon the fact that the concentration of surface active molecules at the interface at time zero will be equal to the concentration of the molecules in solution, which is generally very low. The dynamic surface tension for solvent inks is expected to undergo much smaller changes in dynamic surface tension. The rationale is that the surface tension of pure water is much higher than pure organic solvents, whereas the equilibrium surface tension values for the formulated inks are of the same order of magnitude.

The most important parameters for controlling dynamic surface tension are the concentration, the surface activity, and the diffusion coefficient of the surface active molecules. Surfactants tend to be used at relatively low concentrations because of their surface activity and because of economics. The result is that commercial surfactants generally require several seconds or more to arrive at equilibrium. The lower alcohols, by contrast, exhibit faster dynamic surface tension properties because of their higher diffusion coefficients and the fact that they are used at much higher concentrations.

DYNAMIC DRYING OF THIN FILMS

The dynamic drying of thin films involves a relevant principle of surface chemistry which is called surface tension driven convection and is typified by the Marangoni effect. The results of this principle were first recorded in the biblical

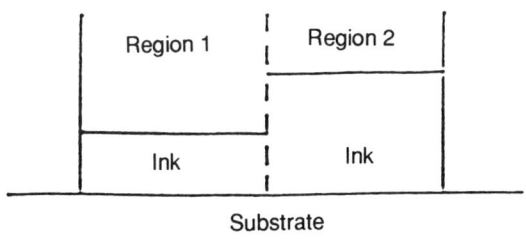

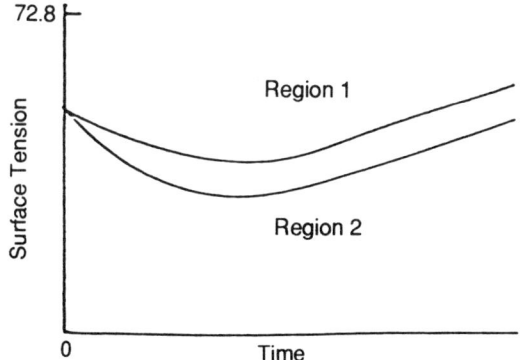

A: Water-based ink containing isoprepanol but no surfactant.

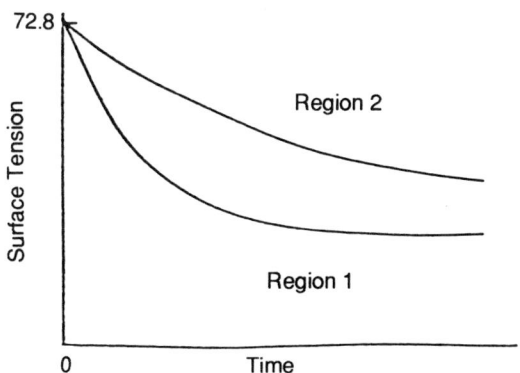

B: Water-based ink containing sufactant and no cosolvent.

Figure 1. Theoretical Prediction of Dynamic Surface Tension for Thin and Thick Ink Films.

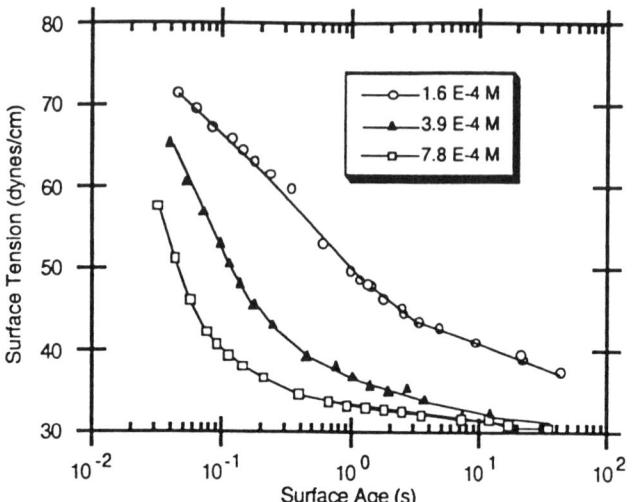

Figure 2. Dynamic Surface Tension of Triton-X100 Solution at Different Concentrations.

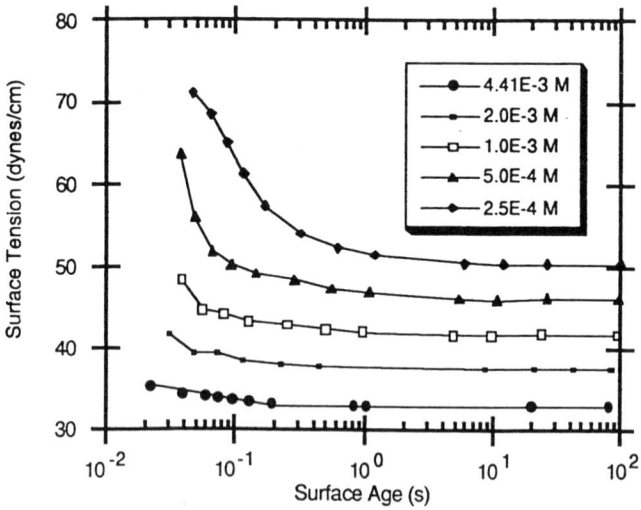

Figure 3. Dynamic Surface Tension of Surfynol-104 Solution at Different Concentrations.

reference to the tears of wine. When wine is rolled in a glass, the ethanol will evaporate at the same rate per unit area from the thin film of wine on the side of the glass as well as from the reservoir of wine in the bottom of the glass. Since the thin film has a higher area to volume ratio than the reservoir, the concentration of ethanol will decrease at a faster rate in the thin film compared to the reservoir. The result will be that the surface tension, which is surface energy, will increase at a faster rate in the thin film. The physical laws of nature require that a system will always tend toward lower energy. The effect of this requirement is that the wine will flow from the low surface tension region of the reservoir up to the higher surface tension region of the thin film in an attempt to reduce the total energy of the system. This is a classic example of surface tension driven convection.

When ink is transferred from a flexographic plate or an engraved anilox cylinder to a nonpermeable surface, such as a film-base material, the initial ink film thickness can be expected to be nonuniform because of the dynamic forces involved in the transfer process. The proposed mechanism for the drying of the ink film is as follows. Adjacent layers of thin and thicker films, which are fluid and free to flow, will initially have the same composition and identical interfacial and surface tensions. Drying will then occur by means of evaporation of solvent from the exposed layers with the result that the ink film will tend towards immobilization due to increased viscosity and eventual solidification. Since the initial evaporation rate in the thin and thicker ink films will be identical, the concentration of surfactant and cosolvent in the ink will vary as a function of time and as a function of ink film thickness over the time frame of fractional seconds from initial ink transfer to final ink immobilization.

Figure 1 presents a prediction of the dynamic surface tension for thin films of water based inks on a nonpermeable surface based on ink composition as a function of time. Region I is defined as the thin ink film and region II as the thicker film, where the two regions are adjacent to one another. Figure 1A represents the predicted dynamic surface tension results for an ink which contains isopropanol as the cosolvent, but no surfactant. Since isopropanol evaporates faster than water the concentration of isopropanol in the ink film will decrease at a faster rate in region I compared to region II. The result will be that after time zero, i.e. the time of ink transfer, the surface tension at any point in time will always be higher in region I than region II. The Marangoni effect or the principle of surface energy driven convection predicts that ink will flow from region II into region I, and the nonuniform ink film thicknesses will be self correcting.

The dynamic surface tension results in Figure 1B are predicted for a water based ink which contains a surfactant and no cosolvent. Since surfactants typically have very low vapor pressures, the evaporation of water will result in the surfactant concentration increasing at a faster rate in region

I compared to region II. Since this is a non-equilibrium condition, the rate of adsorption of surfactant at the ink film surface will be proportional to surfactant concentration, and the surface tension at any point in time after time zero will always be lower in region I than region II. The principle of surface tension driven convection requires that after time zero ink will flow from region I into region II and the result will be that the thin layer becomes thinner. This condition leads to an increase in uniform coverage, i.e. mottling or puddling of the ink film. The proposed mechanism of surface tension driven convection which leads to non-uniform ink coverage or ink mottling has been presented based on the expected dynamic surface tension at the liquid/air interface. The dynamic interfacial tension at the liquid/solid interface, however, is just as important

RESULTS AND DISCUSSION

WETTING OF MODEL GRAVURE INKS

A basic formulation for water based model gravure inks was developed with 7% by weight PCN-Blue and 18% Resin PVP-K30. The variables in the model ink formulation were the concentration of isopropanol as the cosolvent with water at 0% and 10% isopropanol, and the surfactant Triton-X100 or Surfynol-104 at different concentrations. The printed substrates were corona treated and untreated polyethylene films, PE. Table I summarizes the surface energy results for the PE films which were evaluated according to the geometric mean theory as outlined above with water and N-bromonaphthalene as the test fluids.

Table I. Surface Tension, dynes/cm, of Untreated Corona Treated Polyethylene Films

Surface	γ^{d}*	γ^{p}**	γ
PE I (Corona Tr.)	33	10	43
PE II (Untreated)	34	1	35

* γ^{d} = dispersion component of surface tension
** γ^{p} = polar component of surface tension

The spreading coefficients for pure water and the model inks on the untreated and corona treated PE films were calculated from Equation 3, and the surface tension and contact angles of the sample fluids on the PE films. The

equilibrium results are presented in Table II. It should be noted that all the results are negative, and that smaller negative values indicate a greater tendency for spreading.

The spreading coefficient results on PE films in Table II indicate that ink 3 has less of a tendency for spreading on both films than inks 1, 2, and 4. The results also show that these same three inks exhibit a greater tendency for spreading on the corona treated PE I than the untreated PE II. The prediction for printability based on these equilibrium results is that inks 1, 2, and 4 will result in more uniform coverage than ink 1 on both PE I and PE II, and that the ink film coverage on the untreated PE II will in general be only somewhat less uniform than on PE I. This prediction, however, is based on equilibrium results. An important aspect of the interpretation based on dynamic wetting is the result that pure water and a 10% IPA water solution has spreading coefficients of -50 and -20 dynes/cm, respectively, on PE I, and -80 and -30 dynes/cm, respectively, on PE II. The fact is that the results in Table II are equilibrium values, and ink transfer occurs under fast dynamic conditions. It can be expected that in the early stages of a newly generated surface for an ink film, the ink will tend more toward the wetting properties of the pure solvent than of the ink at equilibrium.

Table II. Spreading Coefficients, dynes/cm, of Model Inks and Water on Untreated and Corona Treated Polyethylene Films

Ink	Formulation	Treated PE I	Untreated PE II
1	0.44% Triton-X100	-8.3	-11.4
2	0.44% Triton-X100 (10% IPA)	-7.4	-11.5
3	0.2% Surfynol-104	-12.3	-14.5
4	0.2% Surfynol 104 (10% IPA)	-9.2	-12.5
	Water	-70.0	-84.0
	Water (10% IPA)	-20.0	-30.0

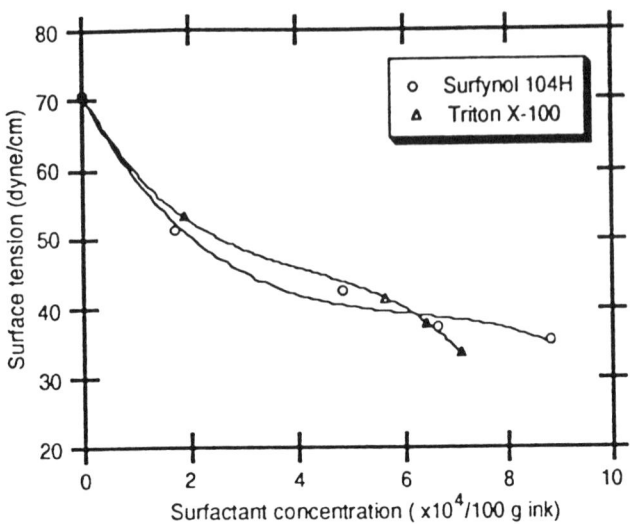

Figure 4. Dynamic Surface Tension of Triton-X100 and Surfynol-104 as a Function of Concentration.

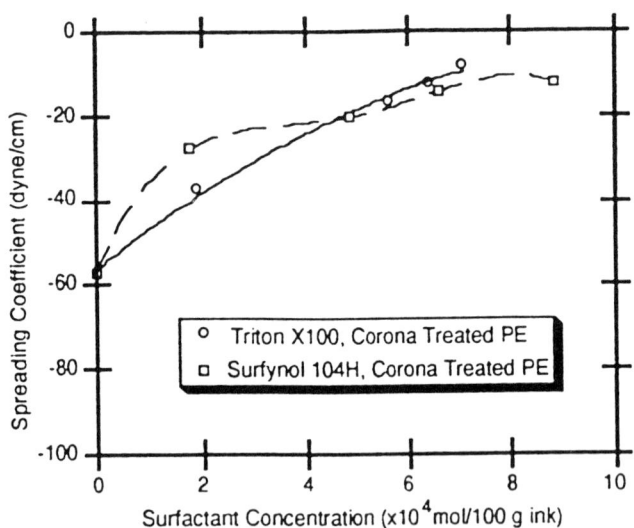

Figure 5. Equilibrium Spreading Coefficient on Treated PE for Triton-X100 and Surfynol-104 as a Function of Concentration.

DYNAMIC SURFACE TENSION OF SURFACTANTS

A drop weight method has been developed for measuring the dynamic surface tension of surfactant solutions at the liquid/gas interface in the time frame of 30 msec to several minutes[6]. The results are presented in Figures 2 and 3 for solutions at different concentrations of the commercial surfactants Triton-X100 and Surfynol 104, respectively. The results show that Triton-X100 exhibits slower dynamics than Surfynol-104 at comparable concentrations. Both of these surfactants have been used in the formulation of the model inks presented in Table II.

The dynamic surface tension results in Figure 2 and 3, which along with the equilibrium surface tension results in Figure 4 and the spreading coefficient results in Figure 5, as a function of the surfactant concentration were used to calculate the dynamic spreading coefficient for the surfactant solutions on both the corona treated and untreated PE films. The procedure adopted was to relate the surface tension values for a given surface age to the surfactant concentration required to achieve the equilibrium surface tension. Figure 5 was then used to determine the spreading coefficient at that surfactant concentration for the given surface age. The assumption is that the rate at which the surfactant molecules arrive at equilibrium at the solid/liquid interface will be the same as at the liquid/gas interface. The calculated dynamic solid/liquid interfacial tension and spreading coefficients for the surfactants used in model inks 1 and 3, Table II, are presented in Figures 6 and 7, respectively. The results reflect the fact that Triton X100 arrives at equilibrium at a much slower rate than Surfynol 104.

LABORATORY INK TRANSFER

The Prufbau Printability Tester, which had been modified to simulate gravure printing[3], was used to execute ink transfer with the model inks on PE I and PE II. A schematic of the modified Prufbau is presented in Figure 8. This instrument is capable of printing in the printing speed range of 80 fpm (feet/minute) to 1100 fpm. An example of the ink transfer results for the model inks 1, 2, 3, and 4 on corona treated and untreated PE film at a printing speed of 300 fpm is presented in Figure 9. The results show that, with the exception of Ink 2 on the untreated PE II and inks 1 and 2 on the corona treated PE I, the remaining model inks exhibit very poor wetting properties with pronounced regions of total de-wetting. These results are in contrast to the spreading coefficient results in Table II which are apparently not capable of predicting the ink transfer results. One example is inks 1 and 2 on PE II where the spreading coefficients are virtually identical and the ink transfer results in Figure 9 are very much different. An important consideration in terms of interpretation of these results, which will be considered in the concluding remarks, is that the results in Table II are based on equilibrium values.

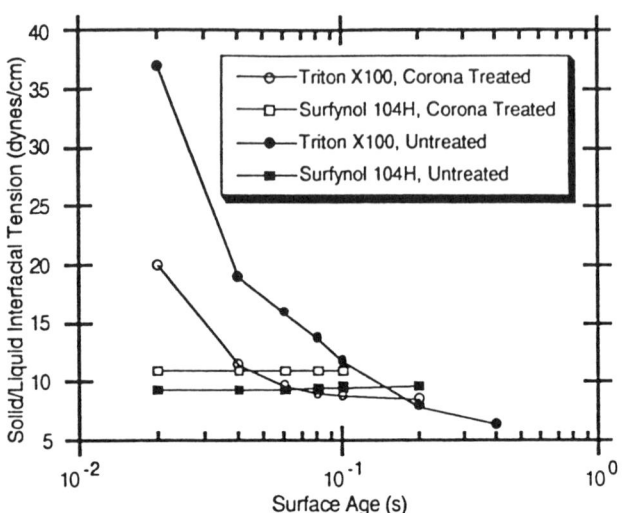

Figure 6. Dynamic Solid/Liquid Interfacial Tension for Triton-X100 and Surfynol-104 on Treated and Untreated PE Films.

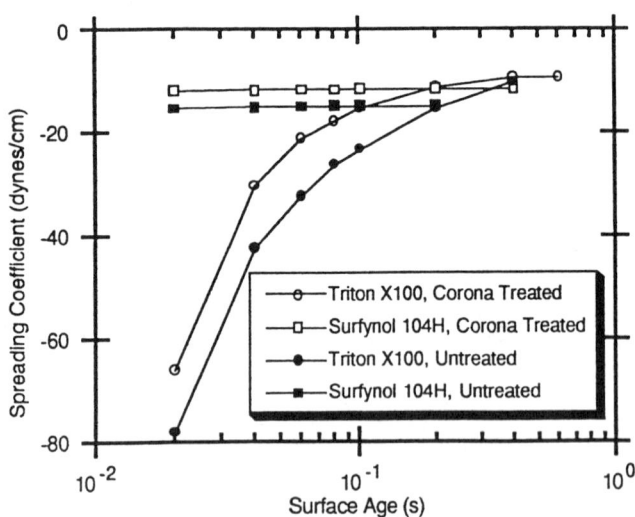

Figure 7. Dynamic Spreading Coefficient for Triton-X100 and Surfynol-104 on Treated and Untreated PE Films.

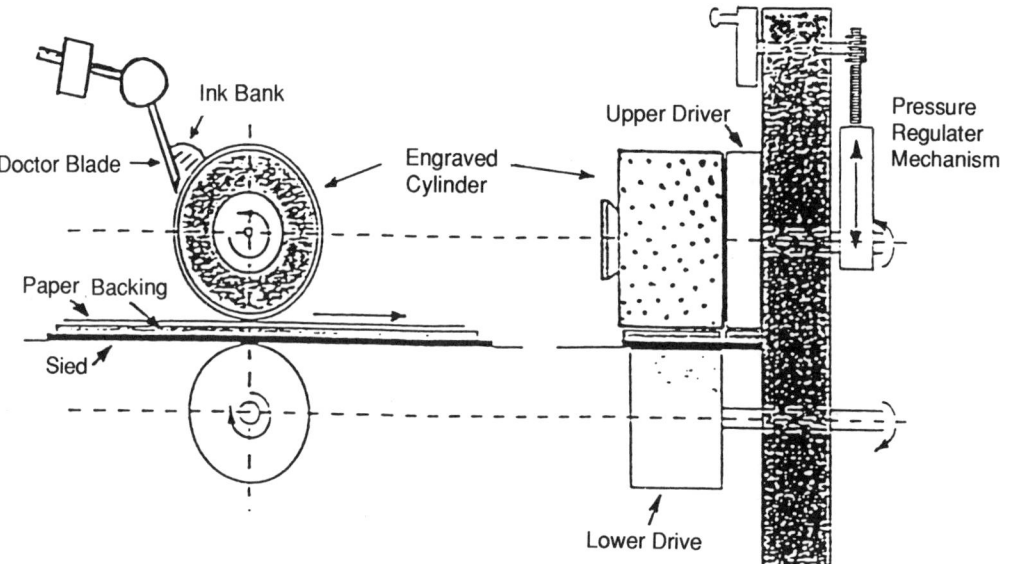

Figure 8. A Schematic Diagram of Modified Prufbau Printability Tester for Laboratory Gravure Printing.

The three inks which did transfer without signs of de-wetting, i.e. inks 1 and 2 on PE I, and ink 2 on PE II, did exhibit varying degrees of ink mottle as determined by the Ultrascan densitometer, but which are not evident from the copies of the prints shown in Figure 9. Ink 2 on PE I resulted in the most uniform print followed by Ink 1 on PE I. The indication is that the addition of 10% IPA to Ink 1 does decrease ink mottle of prints on PE I and improves the wetting properties of Ink 1 on PE II. The addition of 10% IPA to Ink 3, which contains the faster acting dynamic surface tension surfactant Surfynol-104, however, has very little effect on the printability of this model ink on both PE I and PE II.

CONCLUSIONS

The spreading coefficient results in Table II would seem to indicate that water based flexo and gravure inks would print at least as well on an untreated as a corona treated PE film. This conclusion would be contrary to all known experimental evidence as well as the ink transfer results presented in Figure 9. The spreading coefficient results for pure water, however, indicate that the corona treated PE I would be more water receptive than PE II. The interpretation is that the spreading coefficient results for the model inks in Table II are based on equilibrium values, whereas ink transfer occurs under very dynamic conditions. The actual dynamic spreading coefficients for the model inks can be expected at some point in time after ink transfer to be in the range between the equilibrium values for the formulated ink and the values for the pure solvent, i.e. pure water or the 10% IPA solution.

Experimental observations of mottling when a water based ink is printed by either flexo or gravure on a film-base substrate appears to be consistent with the mechanisms involved for surface tension driven convection and dynamic surface tension presented in Figure 1. The result that a water based ink with no cosolvent leads to extensive mottling of the ink film, and that the degree of mottling is decreased with increasing concentration of IPA in the ink formulation is consistent with the concept of surface tension driven convection and the printability results presented in Figure 9 for Ink 1 with Triton X100 as the surfactant. The fact that the same relationship is not maintained for Ink 3 with Surfynol 104 as the surfactant may be related to the fact that this surfactant comes to equilibrium much faster than Triton X-100. An interpretation of dynamic surface tension at the liquid/gas interface, Figures 2 and 3, dynamic surface tension at the solid/liquid interface, Figure 6, and the dynamic spreading coefficient, Figure 7, remains open to speculation. Additional experiments of a definitive nature must be performed in order to more thoroughly understand the dynamics of spreading of thin film on nonpermeable surfaces.

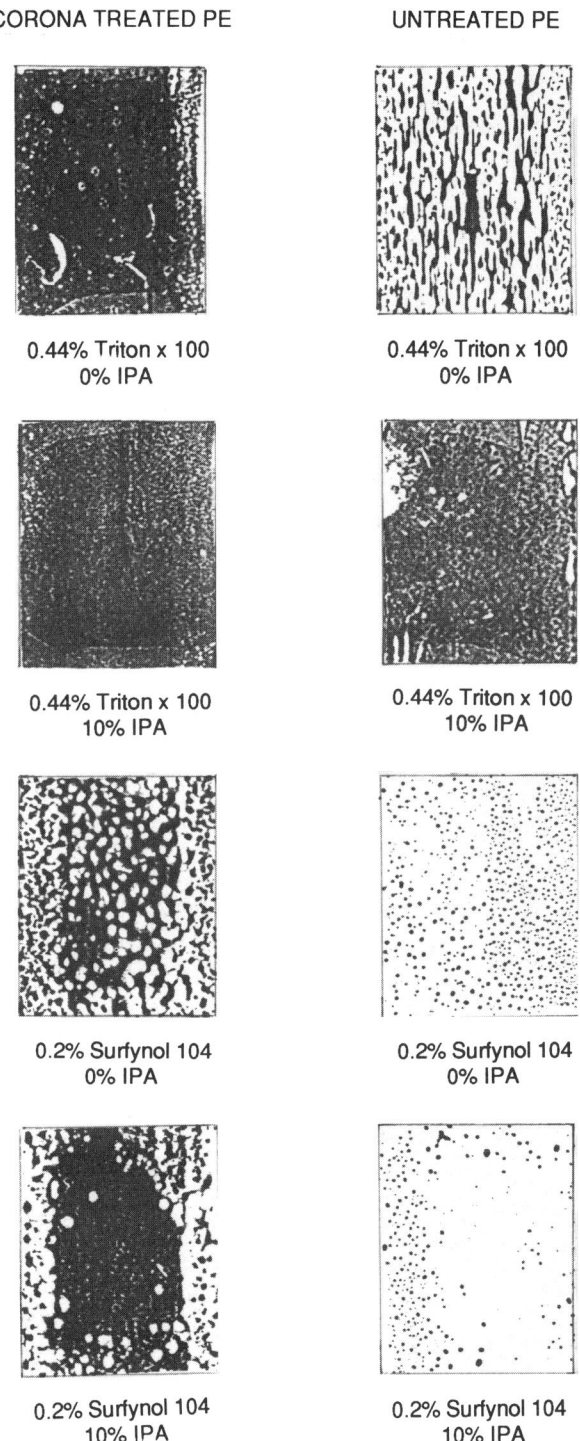

Figure 9. Ink Transfer for Model Inks 1, 2, 3, and 4 on Treated and Untreated PE Films.

REFERENCES

1. Micale, F.J., Iwasa, S., Lavelle, J., Sunday, S. and Fetsko, J., "The Role of Wetting-1", American Ink Maker, **67(9)**, 44-54 (1989).

2. Micale, F.J., Iwasa, S., Lavelle, J., Sunday, S. and Fetsko, J., "The Role of Wetting-2", American Ink Maker, **67(10)**, 25-35 (1989).

3. Lavelle, J., Micale, F.J., and Fetsko, J., "Simulated Printing on the Prufbau Printability Tester", American Ink Maker, **67(8)**, 50-69 (1989).

4. Adamson, A.W., <u>Physical Chemistry of Surfaces</u>, Fifth Ed., John Wiley & Sons, pp 110-115 (1990).

5. Owens, D.K. and Wendt, R.C., "Estimation of the Surface Free Energy of Polymers", J. Appl. Poly. Science, **13**, 1741-1747 (1969).

6. Henderson, D. and Micale, F.J., J. Colloid Interf. Sci., **158**, 289-294 (1993).

COMMERCIAL PRINTING GLOSS

AND PIGMENTS

Arthur C. Rudolph

TRPI
Allendale, New Jersey 07401

ABSTRACT

This paper relates to the influence of pigment properties such as particle size, particle size distribution, dispersion stability of pigments etc. on the gloss of the printing ink films. The dimensions of glossy surfaces are discussed as a classical theoretical model and related to the pigment particle size as measured by Scanning Electron Microscopy (SEM). Fine pigment particles are a major element of glossy printing inks. The following ink-gloss-pigment related topics are presented. (1) Commercial printing-Substrates, (2) Printing ink dimensions and/or Substrate dimensions, (3) Pigment-ink dispersion/Ink additives and (4) Additive nature of gloss. The goal of this presentation is to enhance the focus of research and development in this field to formulate waterborne coatings and printing inks.

INTRODUCTION

Fine pigment particles in printing inks are necessary for several end-use applications of the printed ink films. Finely dispersed organic pigment particles in printing inks are a major contributing factor regarding cost, color strength, color purity, gloss, quality control, printability and physical properties of the coated and/or printed film. A poorly dispersed pigment used in formulating printing ink can cost a lot, not have gloss and/or perform as a quality ink. Generally, gloss is an ink property that is associated with fine pigment dispersions. Commercial printing inks have always utilized fine pigment dispersions. In its simplest form, a printing ink contains pigment and printing vehicle. The

Surface Phenomena and Latexes in Waterborne Coatings and Printing Technologies, Edited by M.K. Sharma, Plenum Press, New York, 1995

139

printing vehicle carries the pigment through a printer and attaches it to an exact area of the substrate. The printer, substrate and required resistive properties dictate the specific selection of pigment and vehicle. Attaining fine pigment dispersions in an array of useful printing vehicles is the technical focus of the printing industry.

EXPERIMENTAL

GLOSS METERS

The printing industry generally uses fixed-angle gloss meters. Two 20° and 60° angles are very popular. The degrees indicate the angle of incidence and reflectance as measured from a normal (90°) to the flat surface. Many feel the 20° angle duplicates the viewing angle of a magazine rack.

PIGMENT-INK-DISPERSION EQUIPMENT

Shear-attrition machines are used to disperse pigments. The pigment-dispersion is facilitated by forcing the pigment-aggregate through a shearing zone. There is an optimum pigment-volume for obtaining commercial quality dispersions from each specific type of dispersion equipment. These ratios are generally established for each color-vehicle-system and considered proprietary information.

Two roller mills are used to produce high quality dispersions in very viscose plastic vehicles. The pigment-vehicle is forced between two rollers pressing together. The pressure is near 100,000 psi and the roller speed ratio is 1.00:1.10. The final solid product is reduced to ink viscosity by addition of lower viscosity vehicle components.

Three roller mills are used to produce paste inks. These inks are produced by passing the paste (wetted pigment-vehicle) through the two nips formed by the rollers. The driving force or roller speed differentiation is 1:2:3. In many cases, a finished lithographic ink is canned directly for sale from this mill.

High speed mixers are used to produce fair to good quality dispersions. Low turbulence disk impellers are used with very high tip speeds (~5000 fpm). In this case, the shear-attrition zone is formed by the tremendous force at the impeller tip and the cooler more viscose ink at the sides and/or baffles. In most cases, inks produced by this method have lower viscosities and utilize a filter operation to remove crude particles.

Knead-shear mixers (Z-mixer, Σ-mixer) are designed to disperse pigment presscake which contains water. The process resembles the action of making bread dough with a mechanical dough hook. Just as flour is worked into the dough, presscake is added to the heavy mixing vehicle. As many as eight additions (bites) maybe required to reach the desired pigment level. At a critical point in the process, the vehicle can replace much of the water on the pigment. This water is drained away and the remainder removed by vacuum if necessary. This method is used to produce high quality flush dispersions for many paste and liquid inks.

Media mills are a popular version of the historic ball mill. Both horizontal and vertical media mills fine spheres as a grinding media. Pegs, disks or screws are used to move the grinding media as the ink is pumped through the grinding chamber. Most mills are cooled and the ground ink is released through a fine slot or mesh panel. These mills are used for a broad range of liquid inks and some news ink.

Kinetic attrition mills are very popular in the water ink area. The impeller and dispersion head are located in the central bottom half of a cylindrical container. Pigment is added to the vehicle pumping in the mill. The path of the pigment is down to the bottom center of the mill and then up through a dynamic pumping zone. Immediately after this area, the pigment goes through a slot and is impinged at high speed on a horizontal surface before beginning the cycle again. Some of these mills are cooled.

RESULTS AND DISCUSSION

REFLECTIVE AND INCIDENT ANGLES

The angle of incidence and reflectance greatly influences the apparent gloss of a flat surface. Greater angles, such as 60°, reflect more light. An elegant treatise on this topic was developed by Fresnel[1] in 1814. The following equations (Table I), express the total reflected light of a perfectly flat surface as the additive total of parallel polarized and perpendicular polarized reflected light.

Table I. Theoretical Relationship Between Reflectance and Angle of Incidence.

FRESNEL REFLECTANCE EQUATIONS

$$R\| = \left(\frac{\cos \theta_i - \sqrt{n^2 - \sin^2 \theta_i}}{\cos \theta_i + \sqrt{n^2 - \sin^2 \theta_i}} \right)^2$$

$$R^\perp = \left(\frac{n^2 \cos \theta_i - \sqrt{n^2 - \sin^2 \theta_i}}{n^2 \cos \theta_i + \sqrt{n^2 - \sin^2 \theta_i}} \right)^2$$

$$R = \tfrac{1}{2} (R\| + R^\perp)$$

$R\|$ = parallel polarized light
$R^\perp$ = perpendicular polarized light
R = unpolarized light
θ_i = angle of incidence
n = refractive index

Figure 1, which follows, is a semi-log graph of the previous equations and illustrates a tremendous increase in reflected light at grazing angles. Fresnel's equation works very well for most ink films printed on flat surfaces. As the equation implies, inks with higher values of n (refractive index) will generally reflect more light. An example is an opaque ink containing titanium dioxide pigment (n = 2.70), which would add greatly to the vehicle refractive index (n=~1.50)

The refractive index of a reflective surface will dramatically effect the apparent gloss. The gloss of a pure white surface, such as titanium dioxide pigment (n = 2.7), would be much higher than the vehicle (n = 1.5), as example given in Figure 1. The difference in gloss at 20° angle of incidence for the two examples is calculated to be 21.1% and 4.1%.

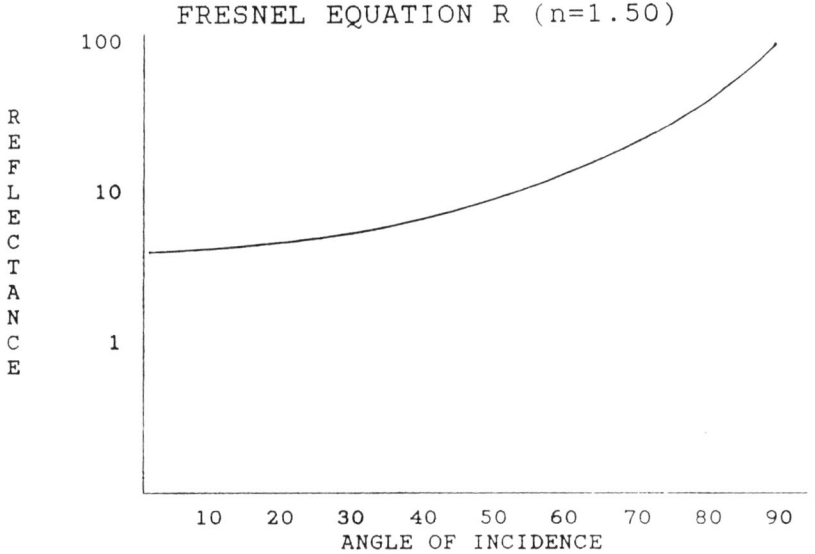

Figure 1. Reflectance as a Function of Angle of Incidence.

A graphic representation of this model as it applies to visible green light (0.550 μm) is presented in Figure 2. Empirical work cited by L. A. Simpson[3] and generated by Bennett and Porteous generally agree with Figure 2. •In their work, An electro-mechanical stylus device (profilimeter) was used to measure surface roughness of paints. The work indicated that 0.07 μm surface defects would not be detected by 20° Gloss meter readings of the less than perfect surface.

SURFACE SMOOTHNESS

This topic has also been of prime concern to a Nobel prize winner, J. W. Rayleigh[2]. In his work he defined the largest surface defect height which did not affect gloss. This height is defined as the wavelength of light divided by 8 times the cosine of the angle of incidence. This is only true for defect heights less than the wavelength of light and is expressed by the following equation:

$$D_h = \frac{\text{Wave Length}}{8 \, \text{Cos} \, \theta_i} \tag{1}$$

PARTICLE SIZE DISTRIBUTION

A scanning electron-beam micrograph was used to establish a gloss-particle size relationship that has been developed thus far. Figure 3 is a micrograph of a glossy dispersion of titanium dioxide. The dispersion was applied by a smooth rubber roller to aluminum foil and dried at ambient conditions.

Figure 4 was constructed to convey the idea that 0.15 μm radius particles are very glossy and 1.10 μm radius particles are non-glossy. In order to study the larger particle size of exactly the same batch of pigment, we mixed the 75% white with latex. Clumps of pigment larger than 2.0 μm were formed. Particle size data from SEM was matched with glossmeter readings to construct Figure 4.

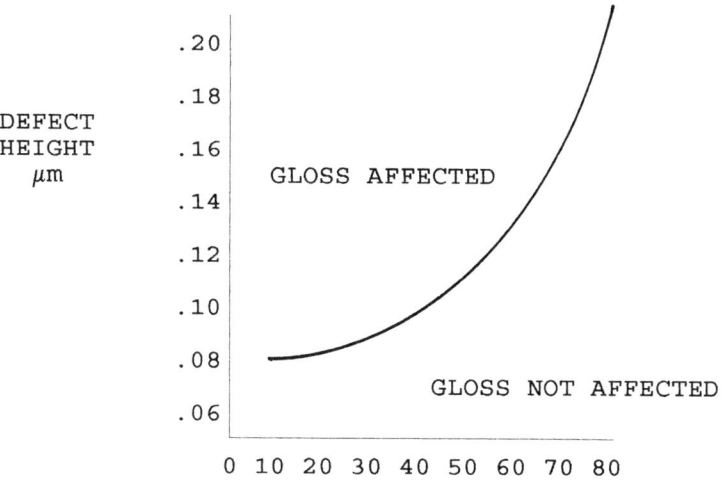

Figure 2. Rayleigh Defect Model (0.550 μm).

INK FILM THICKNESS

It should be noted that in very thin coatings (0-0.7 μm) extreme variations in reflectance may occur. This phenomenon is exhibited by differences in substrate refractive index and coating refractive index[4]. Greater reflectance differences happen at thicknesses between 0.0-0.5 and 0.5-1.0 μm. Ink films we are considering in this paper are greater than 2.0 μm and will not be greatly effected by this phenomenon.

20° GLOSS - 80.0%   1 μm

Figure 3. A SEM picture of a glossy dispersion of TiO₂
(15000X, 75% TiO₂)

Surface roughness is a common and effective way to express lack of gloss. In sanding wood coarse paper causes large and deep defects and results in very low gloss. Sanding again with a fine sand paper will make the surface defects smaller and less deep. The result of this change is an increase in gloss. Figure 4 indicates that very small defects (1.0 μm) will dramatically depress gloss.

GLOSS AND SUBSTRATES

Substrates fall into two distinct groups. Porous substrates, such a papers, boards, woven and non-woven fibers, are one. The other group, non-porous substrates, includes foils, cast and extruded plastics. Penetration of ink vehicle into the substrate is greatly depressed when printing foil, as an example. Gloss in this case is greatly enhanced by the presence of vehicle to minimize the negative effects of critically large pigment particles. In the case of paper, ink vehicle will be absorbed and leave a quantity of pigment on the surface. The amount of pigment and it's effect upon gloss will depend upon the substrate as well as the quality and stability of the pigment dispersion. Chrome coat paper is an example of a very glossy porous paper which requires high quality inks to maintain gloss because of vehicle absorption.

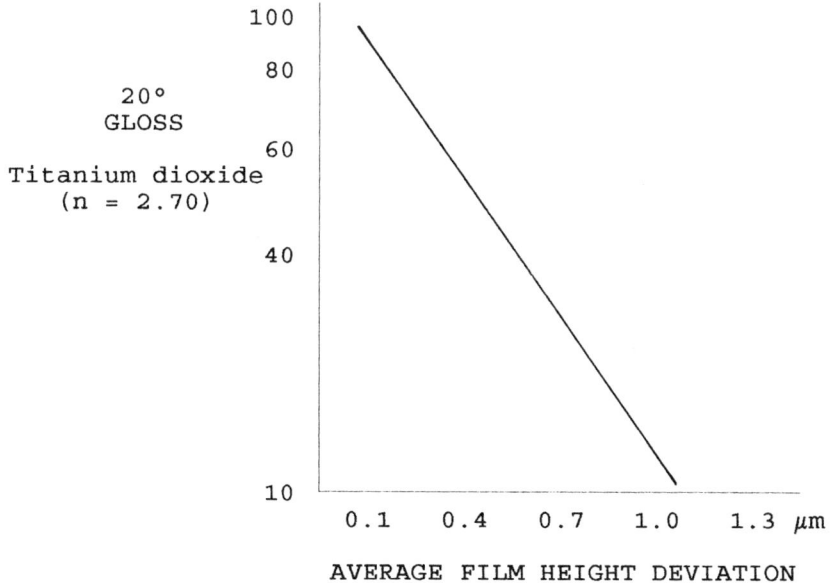

Figure 4. Film Gloss on a Non-Smooth Flat Surface Versus Film Thickness

PRINTING INK DIMENSION AND RHEOLOGY

The thickness of the printed ink film and the inks resistance to flow can influence gloss. Table II is a general description of thickness of different printing methods and flow. Flow is measured in poise at 25°C.

Table II. Ink Film Thickness and Viscosity of the Printing Inks Used in Different Printing Processes

PRINTING METHOD	INK THICKNESS	POISE
Gravure	15 μm	0.30
Flexography	8 μm	1.50
Letterpress	6 μm	50.00
Lithography	3 μm	150.00

The lower poise systems, such as gravure and flexo, have generally poorer gloss on porous substrates. In these systems the vehicle is efficiently absorbed by the substrate. In this case, critically large particles are exposed and depress gloss. However, there is a distinct advantage in using these systems on non-porous substrates. The good flow of these systems helps maintain the gloss of the substrate. Printing ink gloss-defects are minimized by the flow. By contrast letterpress and lithography enhance gloss on papers and board, and are more critical on foils and plastic substrates.

SUBSTRATE DIMENSIONS

The dimensions of porous substrates, such as coated and un-coated paper, are very large in relationship to ink thickness. Some examples[5] are presented in Table III.

COMMERCIAL PRINTING

The following schematics are caricature of the printing processes (Figures 5-8). They are included to convey the differences in the ink distribution systems. These distribution systems are designed for high and low viscosity inks. The simpler systems (gravure and flexo) use low viscosity fluid inks (Figure 5, 6). The multi-roller systems (Letterpress and Litho) exist to spread out high viscosity paste inks (Figures 7 and 8).

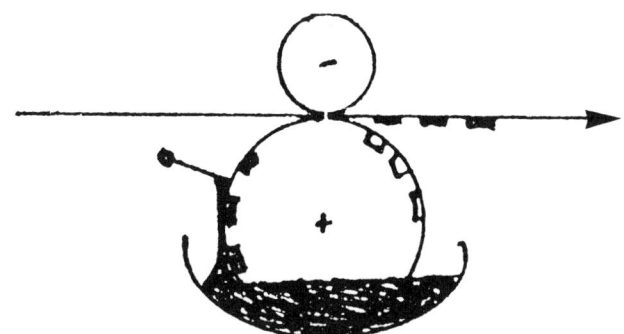

Figure 5. A Schematic Illustration of Gravure Printing Process.

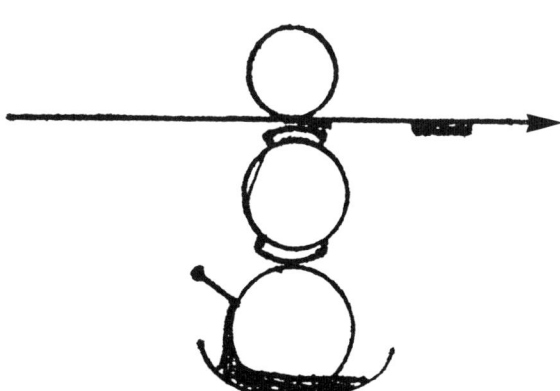

Figure 6. A Schematic Illustration of Flexo Printing Process.

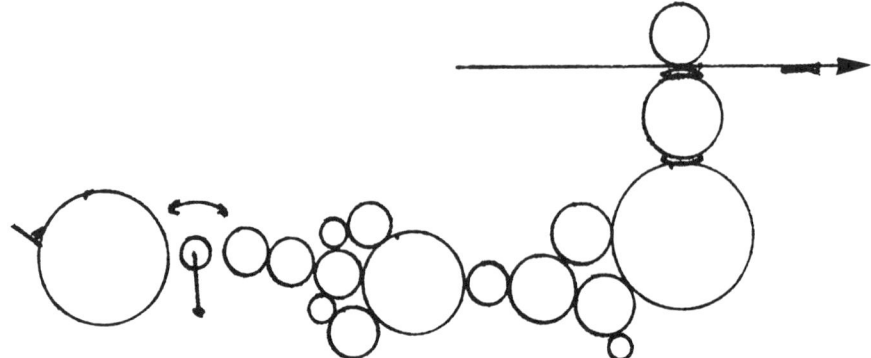

Figure 7. A Schematic Illustration of Letter Press.

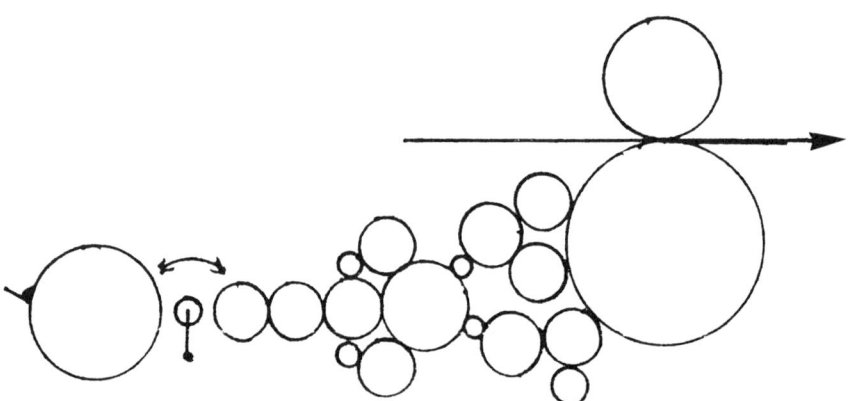

Figure 8. A Schematic Illustration of Litho Press.

Table III. Dimensions of Porous Substrates Used in Printing Processes.

SPECIMENS(70# CIS TEST PAPER)	UNSUPER CALENDERED	SUPER CALENDERED	QUALITY SHEET
Hole Count	10	4	2
Hole Diameter (μm)	40	20	15
Fiber Diameter (μm)	10-25	8-25	2-25
Cross Section: Top to Bottom Surface Valley Measurements	40	25	10

SURFACE DEFECTS/FIELD 0.20 sq.mm.

PIGMENT VEHICLE INTERACTIONS

Very small, stable pigment particles in a suitable vehicle is the goal of an ink chemist. The interactions in the ideal ink model allows the wetting and coating of the pigment particles within the normal clusters of pigment particles to attain a stable pigment dispersion. The word stable in the last sentence indicates a fine division of pigment particles that will not re-cluster (flocculate) when shocked by formulation additions or printing.

An example of an ideal gravure printing ink would be the dispersion of AAOA diarylide yellow in a vehicle composed of α-methyl polystyrene solvated in toluene. In this case, there is a commonality of chemical moieties in the pigment, polymer and solvent. This ideal ink required very little milling and gave a high quality stable glossy transparent dispersion. The unique aspect of this ink is that α-methyl polystyrene is a poor dispersing polymer for most pigments.

In many printing inks, there are barriers which inhibit the production of fine stable pigment dispersions. These barriers are presented by chemical-physical differences between ink components. In a practical sense, dispersants are used to minimize these differences by introducing compatibility bridges between the components. These bridges enhance the dispersion and stability of fine pigment particles through the ink vehicle. The end effect desired is less flocculation which improves color value and gloss. Clearly defining pigment surfaces is a major goal in this work. To further complex the problem, many pigments have chemically different faces. This suggests that optimum use of these pigments may require multiple surfactants.

Hyper-dispersants are an effective approach to attaining fine stable pigment dispersions in difficult pigment-vehicle systems. A hyper-dispersant has two segments. One alines with a pigment face and the other enhances the coating of the pigment with a stabilizing vehicle. Many of the hyper-dispersants are low molecular weight materials and do not contribute greatly to flow during the dispersion process. The understanding of hyper-dispersants for specific pigment faces is a major contribution to our industry.

INK ADDITIVES AND GLOSS

Ink additives can generally decrease gloss if they are not soluble in the vehicle. Wax like materials (powdered polyethylene) are added in small quantities to ink. The purpose is to enhance rub and scratch resistance. The size of the polyethylene particles cause a decrease in gloss. The amount of the material added is critical regarding gloss and rub resistance properties and is tightly controlled.

CONCLUSIONS

Glossy printing is dependent upon ink, ink-curing and substrate. All of these parameters affect the final surface presented for gloss evaluation. Our work has indicated that surface defect heights greater than ~0.1 μm can depress gloss. The ink-gloss effect is controlled by pigment dispersion, ink film thickness and ink penetration. In the ink, the size of the pigment aggregate is critical regarding the smooth surface required for gloss. Generally, pigment aggregates, pigment particles and other particles larger than the ink film thickness will detract from a glossy surface. Large particles maybe found or formed in both ink and substrate during the printing process. However, a relatively glossy print must use a glossy ink. This glossy ink on the print surface must contain a fine stable pigment dispersion.

The major technical goal of a printing ink developers and their suppliers is to attain a fine stable pigment dispersion in suitable printing vehicles. New printing vehicles are developed every month and inherently have some pigment-dispersion problems. There is much to gain by developing unique pigment-dispersant-vehicle systems. Enhanced pigment dispersions give greater gloss, better color value, and improved printability. In addition, quality dispersions improve cost and waste factors resulting in less environmental impact. This paper suggests that our Industry should focus upon the optimum use of pigments for these benefits. We feel our Industry generally wastes color and should develop or solicit technical support. The Industry and our Society will benefit greatly from these future technical advances.

KEY WORDS

Commercial printing   Printing inks
Flocculation   Printing Precesses
Gloss elements   Printing Substrates
Pigment dimensions   Refractive Index
Pigment dispersions   Surface Smoothness

REFERENCES

1. Ditchburn, R.W., LIGHT, Blackie & Son, London, pp530, (1963).

2. Rayleigh, L., NATURE, **64**, 385, (1902).

3. Simpson, L.A., PROGRESS IN ORGANIC COATINGS, **6**, 1-30, (1978).

4. Heavens, O.S., OPTICAL PROPERTIES OF THIN SOLID FILMS, Academic Press Inc., pp156-157,

5. Rudolph, A., RADCURE EUROPE CONFERENCE 1985, FC85-431, Society of Manufacturing Engineers, Dearborn,MI 48121, (1958).

ELECTROKINETIC PROPERTIES OF POLYMER

MACRO-SURFACES

V. Ribitsch

Department of Physical Chemistry
University of Graz
Heinrichstr. 28
A-8010   Graz
Austria

ABSTRACT

The surface charge of polymer macro-surfaces is a critical parameter for many technical processes. This important material property of films, fibers and granulates can be determined employing streaming potential measurements. The application of this method to irregular formed material is presented. The described method is suitable to characterize the surface of solids by two methods. The isoelectric point can be determined by measuring the pH dependency of the zeta potential and in consequence the pK values of surface groups can be estimated. The number of charged groups and adsorption properties can be determined by measuring the zeta potential as a function of the surfactant concentration in the liquid phase. In consequence either the solid can be characterized with the help of well known surfactants or surfactants can be characterized using well known solids as adsorption matrix. Examples of measurements are presented to demonstrate the use of the method for basic research as well as for the investigation of several important industrial applications.

Surface Phenomena and Latexes in Waterborne Coatings and Printing Technologies, Edited by M.K. Sharma, Plenum Press, New York, 1995

153

INTRODUCTION

Electrokinetic surface properties of materials are important for a number of industrial processes as well as for biological and medical tasks i.e. the performance of artificial vessels[1,2]. These properties are generated by the electrochemical double layer described by Guy, Chapman, Stern and Graham which exists at phase boundaries between solids and electrolyte solutions and described by the zeta potential (ZP)[3].

The following assumptions according the origin of the double layer and the calculation of the ZP are made :

> Surface charges are due to dissociation of chemical groups and or preferential adsorption of ions or adsorption of ions in the inner Hellmholtz plane. The zeta potential = potential at the outer Hellmholtz plane Validity of the Hellmholtz-Smoluchowski rule

The zeta potential can be calculated by the well known Smoluchowski equation from streaming current (Is) or streaming potential (Us) data[4,5].

$$= \frac{Is' \eta' L}{dp \, \varepsilon \, \varepsilon_0 Q} \qquad (1)$$

In order to determine the zeta potential of coating and printing films, as well as fibers correctly, it is necessary to measure the streaming current/streaming potential as a function of the pressure difference (dp) in the electrolyte solution between cell inlet and outlet. The term Is/dp represents the slope of the line if Us is plotted vs dp. The offset describes the asymmetric potential of the electrodes and does not influence the results. The geometrical term L/Q (length over cross section of the channels between the fiber) has to be determined and this causes some problems.

Several methods were described[6] to determine the quotient L/Q which describes the capillary system in the fiber plug. We used the method of Fairbrother and Mastin (F/M), where the quotient is determined by measuring the resistance along the cell (fiber plug).

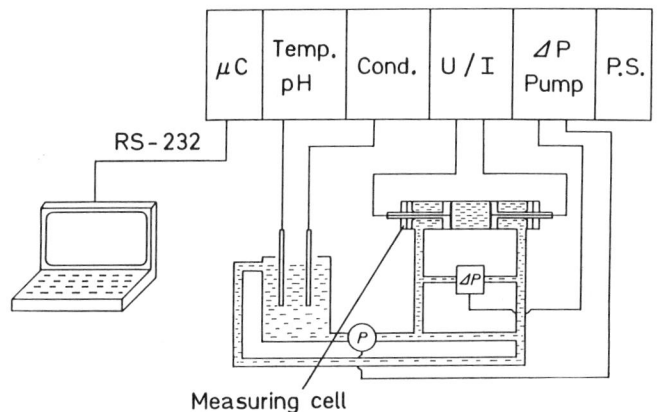

Figure 1. A Schematic Illustration of Electrokinetic Analyzer EKA.

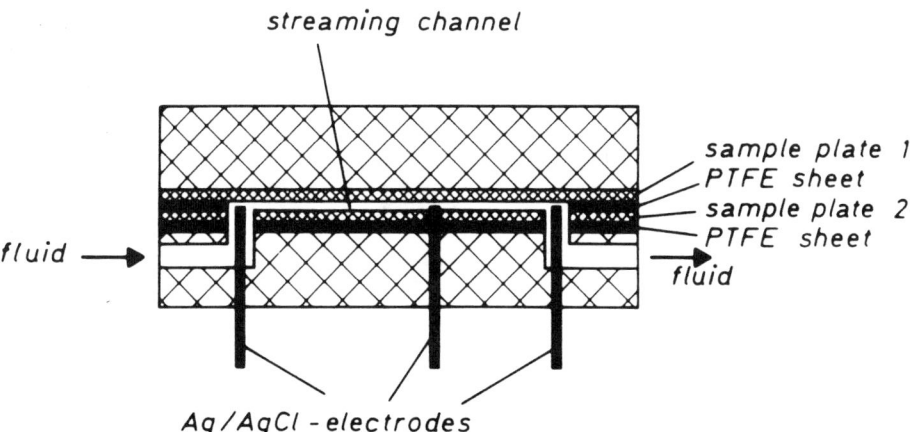

Figure 2. Fiber Cell.

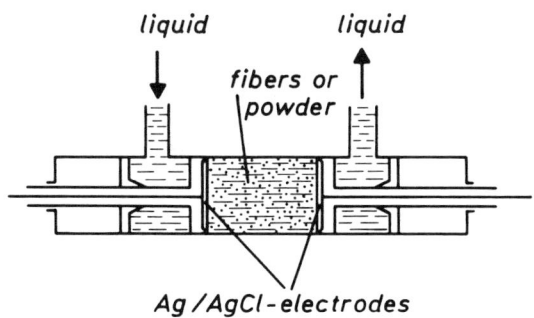

Figure 3. Film Cell.

EXPERIMENTAL

An instrument[7] was developed (Electrokinetic Analyzer EKA, Figure 1) for measuring the streaming potential/ streaming current of fibers, granulates and coating films (flat surfaces) according to two accepted methods (Fairbrother and Mastin resp. Chang and Robertson[8]).

The Electrokinetic Analyzer EKA[5] is a modular and computerized system which works in the following way: An electrolyte solution (10^{-3} M KCl) is forced at controlled increasing pressure through a bundle of capillaries (plug) or a small channel built by two flat surfaces. The streaming current (Is) or the streaming potential (Us) resulting from the motion of ions in the diffuse layer is measured.

The fiber cell (Figure 2) consists of two perforated sliding Ag/AgCl electrodes. The fibers are placed between these electrodes, compressed and the electrolyte solution is forced through the plug. Granulates can also be packed between the electrodes of the fiber cell with the help of special adapters in order to detect the ZP according to equation 1.

Parallel plates and films are investigated in a cell creating a channel of well defined distance between two sample surfaces (Figure 3). The ZP is calculated according to the method of Fairbrother/Mastin. Irregular formed probes can be placed into sealing material (i.e. wax); coating and biological material such as polymer films, skin and vessel walls was measured by this method.

With the help of this method the ZP of almost every kind of solid as a function of pH, electrolyte concentration, different electrolyte components, as for example surface active substances, can be studied.

RESULTS AND DISCUSSION

The effect of several parameters such as zeta potential, pK, IEP etc. versus pH was examined for polymers in order to characterize the polymer surfaces.

CHARACTERIZATION OF POLYMER SURFACES

If dissociating groups are present at the surface, the following can be assumed:
1. The alteration of the ZP as a function of pH is due to the dissociation of surface groups.

2. The plateau of the ZP - pH curve is due to complete dissociation of functional groups.

The pH dependency of the ZP for several polymers is shown in Figure 4. Total dissociation is assumed at the plateau regions. The maximum in zeta potential was observed at the plateau, which correlates well with the hydrophilic

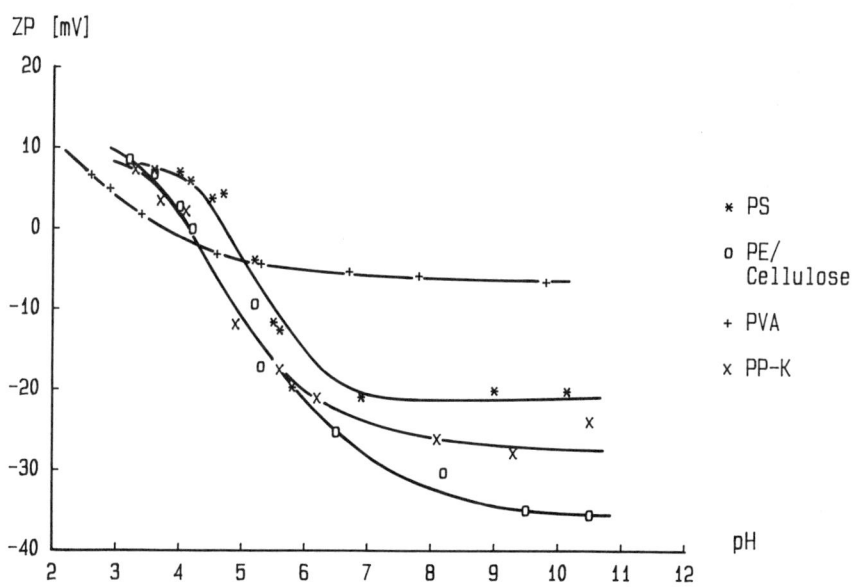

Figure 4. Zeta Potential - pH Curves of Several Polymers.

Acid groups :

$$HA^- + H_2O \rightleftharpoons H_2O^+ + A^-$$

$$K_A = \frac{[H_3O]_s^+ \ [A-]^-}{[HA^-] \ [H_2O]}$$

$$[H_3O]_s = [H_3O^+]_b \ \exp\left(-\frac{F\zeta}{RT}\right)$$

$$\log K_A = pK_A = pH + 0.4343 \left[\frac{F\zeta}{RT} + \ln\left(\frac{\sinh\left(-\dfrac{F\zeta_{plateu}}{2\,RT}\right) - 1}{\sinh\left(-\dfrac{F\zeta}{RT}\right)} \right) \right]$$

Basic groups :

$$-B + H_2O \rightleftharpoons BH^+ + OH^-$$

$$K_B = \frac{[BH^+] \ [OH^-]_s}{[B] \ [H_2O]} = \frac{[BH^+] \ K_W}{[B] \ [H^+]_s}$$

$$[OH^-]_s = [OH^-]_b \ \exp\left(\frac{F\zeta}{RT}\right)$$

$$\log K_B = pK_B = pK_W - pK_A$$

b	=	bulk
s	=	surface
F	=	Faraday constant
R	=	Gas constant
T	=	Temperature
K_W	=	Ionic product of water

Figure 5. Equations to Determine the Dissociation Constant (pK Value) From the Zeta Potential - pH Function (Boerner 1987).

properties of the polymer. The maximum in zeta potential and the isoelectric point (IEP) are characteristic surface parameters of a polymer.

Under the assumption of the GCSG model Boerner[9] has developed a method (Figure 5) to determine the dissociation constant from the ZP - pH function. The following assumptions were made for this approach :

> zeta potential = potential at the outer Helmholtz plane adsorption of ions in the inner Helmholtz plane the capacity of the double layer is smaller then 25 umF/cm².

Carboxyl groups are the origin of pK values of 4.5, higher pK values are due to the additional adsorption of OH⁻ ions.

ZETA POTENTIAL VERSUS pH FOR PRINTING PLATES

Aluminium sheets are used in offset printing (Figure 6) and undergo a very specific surface treatment in order to perform their task. Figure 11 shows the pH dependency of the ZP of different treated Aluminium surfaces, one called 'dim' the other one 'metallic'. Both examples show the pH - ZP curves typical for acid groups, but due to the surface treatment they exhibit completely different ZP - pH functions. The maximum in zeta potential (hydrophilic) of the dim sample is clearly higher then that of the metallic one, the IEP differ at least two units of the pH scale. Interactions of charged groups with these surfaces will, of course, depend on the hydrophilic properties as well as the applied pH because the surface will exhibit totally different behavior.

Table I. IEP - pK values of polymer materials calculated from zeta potential - pH functions.

POLYMER	IEP	pK
PE/Cellulose (50/50)	3.9	5.0
PS	4.9	5.2
PP	3.9	4.9
PP coron.	3.7	4.5
PVA	3.7	4.5

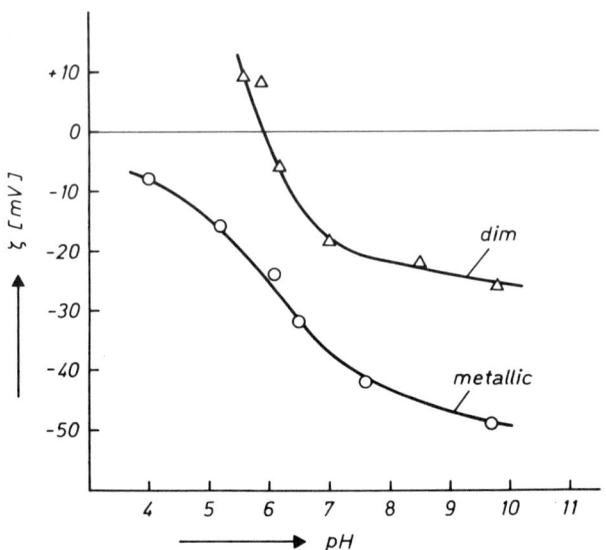

Figure 6. Zeta Potential as a Function of pH of Offset Printing Plates.

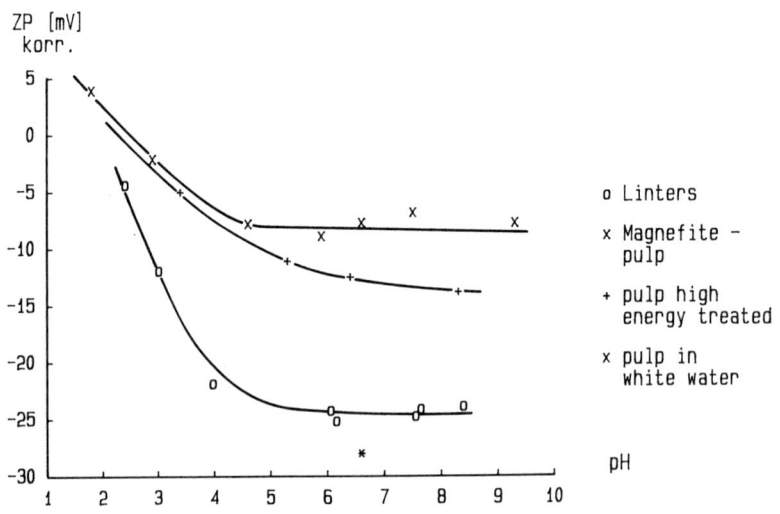

o Linters

x Magnefite – pulp

+ pulp high energy treated

x pulp in white water

Figure 7. Zeta Potential Versus pH of Different Cellulose Fibers (Pulp).

The hydrophilic cellulose fiber (magnefite-pulp) shows (Figure 7) constant low zeta potential of - 7 mV between pH 4.5 and 9 which decreases sharply at pH values below 4 leading to an IEP around 2.5. Linters exhibits a very similar pH function except that the maximum in zeta potential is around -25mV. The constant ZP at pH values above 5 is caused by the total dissociation of carboxyl groups. A treatment of the magnefite - pulp with high energy pulses (in order to improve the reactivity for grafting processes) changes dramatically the ZP-pH curves. The plateau is shifted and almost disappeared. If the electrolyte solution is changed against the 'white water' used in paper processing, the ZP (at pH 6) of -28 mV is observed. This indicates the adsorption of so called anionic trash (resin components) on the fiber surface.

The surface charge of composites as paper can also be measured (Figure 8). Uncoated papers show very similar ZP - pH curves. A plateau of positive ZP is observed at pH value > 6. Differences in the absolute values of the ZP are caused by different fillers.

Coated papers show completely different behavior depending on the type of coating pigment and polymer additives. The carbonate coated sample was manufactured in an alkaline environment, a plateau above pH 4.7 and an IEP at pH 4.2 are observed. The kaolin coated sample exhibits an almost constant negative ZP of - 10 mV. This is in disagreement to the ZP values obtained from the pigments itself at the same pH values. The results reflect the influence of polymer additives, surfactants and latex on the surface charge. A correlation between ZP and printability of the papers are discussed but there are not enough data available to draw any conclusion in detail at this moment.

INVESTIGATION OF SORPTION PROCESS

The interaction of the solid with components of the solvent (adsorption, desorption) depends on the sign of charge of both components. Adsorption of species with the same charge as the solid phase cause an increase of the ZP whereas species of an opposite charge decrease the ZP, finally leading to a charge reversal at sufficient concentration.

This phenomenon can also be used to characterize surfaces which do not show differences in the chemical nature such as the amount of dissociating groups or adsorption site.

As an example the adsorption of cetylpyridinuim chloride (CP) on polyacrylnitril fibers is shown on Figure 9. The cationic surfactant is added into the electrolyte solution and the ZP is measured as a function of surfactant concentration. The adsorption of the cationic surfactant on the anionic fiber causes a decrease of the negative ZP. A concentration of 1.7 mM CPC is necessary to obtain the zero point of charge (ZPC). With the addition of more cationic surfactant, the ZP and hence the sign of the surface charge of the fibers is changed into a positive one.

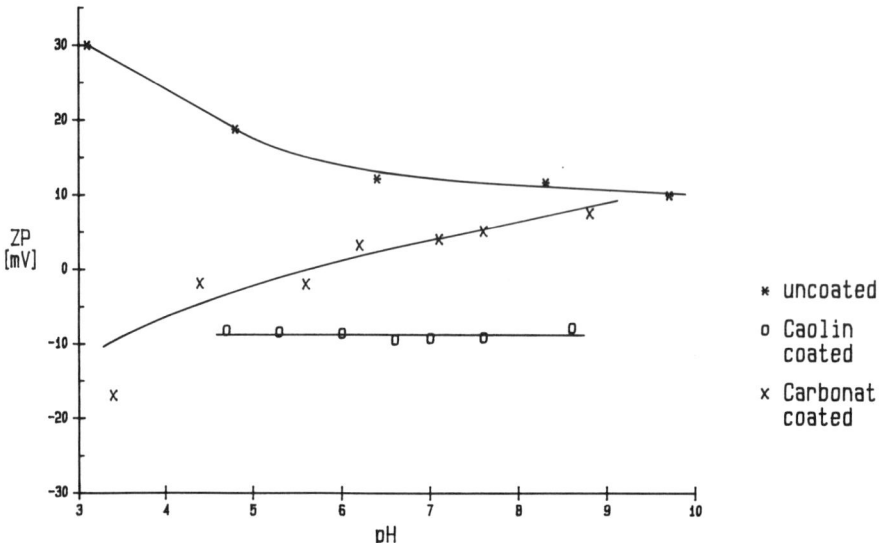

Figure 8. Zeta Potential Versus pH of Different Papers.

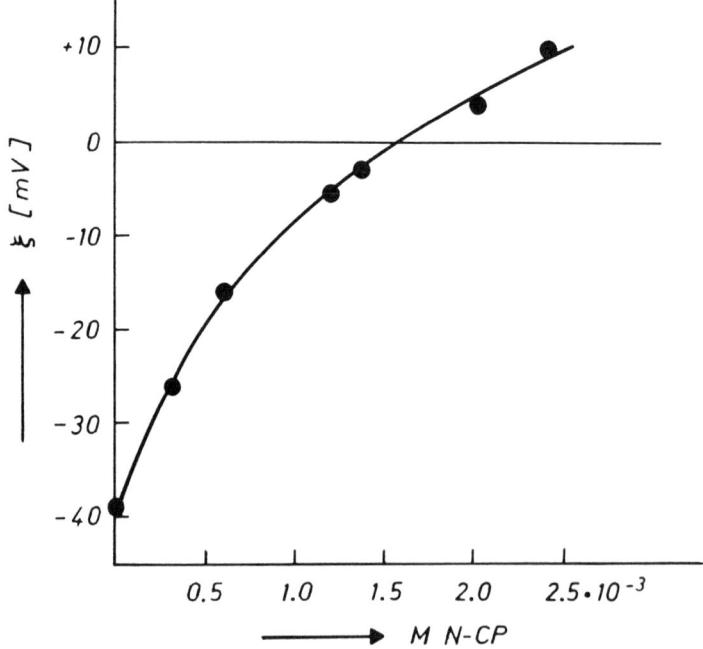

Figure 9. Adsorption of Cetyl Pyridinium Chloride (CPC) on Polyacrylnitril Fibers.

INDUSTRIAL APPLICATIONS

Polyethylene/cellulose filters can be investigated in both cell types, the fiber and the film cell. The results agree very well. The IEP do not differ if the concentration of the compounds is changed. But the amount of cationic surfactant to obtain ZPC is different as shown in Table II. The difference in the compounds of the two types of filters can be characterized in this way. These methods can also be used to determine the capacity of ionic filters.

The adsorption of surfactant (Figure 10) can, for example, be used to study the result of a treatment of polypropylene with high energy pulses in order to obtain a printable surface (increase the adhesion of paint and the wettability - hydrophilicity). The irradiated PP-film does not show significant differences in the IEP and the ZP - pH function compared to the untreated sample.

Contrary to the almost identical ZP - pH function the adsorption of surfactants is characteristically different. The surfactant concentration (CPC) necessary to achieve the ZPC is more then twice compared to that of the untreated sample. This indicates an increase of the number of adsorption places by the corona treatment. The difference of the surfactant concentration necessary to obtain the ZPC and it can be used to characterize differences in polymer surfaces.

Table II. Filter material - IEP and Cetypyridinium chloride to obtain ZPC.

FILTER COMPOSITION (POLYESTER/ CELLULOSE)	FILM-CELL IEP (pH)	FIBER-CELL IEP (pH)	CPC(mg)/ FIBER(g) TO OBTAIN ZPC
90/10	4.2	4.2	12.9
70/30	3.9	3.9	-
50/50	3.9	-	7.4

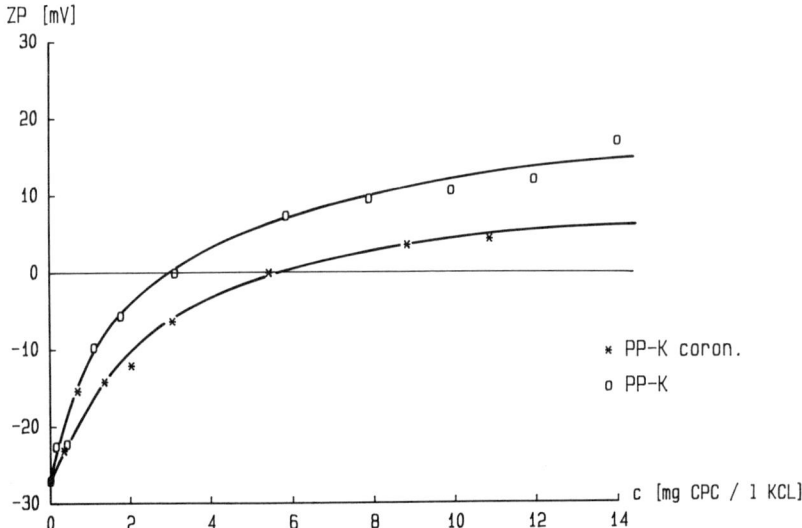

Figure 10. Zeta Potential of Polypropylene Versus Surfactant Concentration.

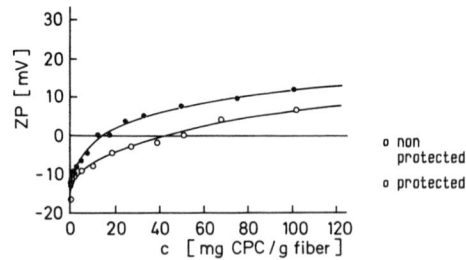

Figure 11. Zeta Potential of Viscose Fibers Versus CPC Concentration.

An other application is the monitoring of the adsorption of coatings on fibers. Viscose fibers, for example, are protected against oxidation by a coating process (Figure 11). The coating does not change the ZP - pH functions of these fibers, they are almost identical.

The amount of surfactant necessary to obtain the ZPC discriminates the quality of the treatment. In the case of oxygenation protected fiber a three fold quantity (12.4 mg/g fiber) of cationic surfactant compared to the unprotected fiber (41 mg/g fiber) in the electrolyte solution is necessary to obtain the ZPC.

CONCLUSIONS

Electrokinetic properties of macro - surfaces can be characterized applying streaming potential measurements. The calculated zeta potential can be used either to characterize the chemical nature of polymer surfaces or of surface active substances. This can, on the one hand, be done by altering the pH of the electrolyte to measure the ZP - pH function and to obtain as characteristic parameters like maximum zeta potential and the IEP. On the other hand, the adsorption properties of ions and surfactants can be investigated to obtain either information about the substrate or the dependency of the adsorption process on a number of parameters i.e. molecular structure, molecular weight of the surfactant and counter ions.

REFERENCES

1. Jacobasch, H.J. and Bauböck, G., Schurz, J. Coll. & Polymer Sci., **3**, 263 (1985).

2. Ribitsch, V., Jacobasch, H.J. and Boerner M., in: Advances in Measurement and Control of Colloidal Processes, Butterworth-Heinemann ,p354, (1991).

3. Duchin, S.S., in: Surface and Colloid Science, Academic Press, New York (1974).

4. Smoluchowski, V.M., Bull. Intern Acad Sci Cracovie, p184 (1903).

5. Hidalgo-Alvarez, R., Advances in Colloid & Polymer Sci., **34**, 217 (1991).

6. Fairbrother, F. and Mastin, H., J. Chem. Soc., **75**, 2318 (1924).

7. Ribitsch, V., Ch. Jorde, Ch., Schurz, J. and H.J. Jacobasch, H.J., Progr. Colloid & Polymer Sci., **77**, 49 (1988).

8. Chang, M. and Robertson, A., Can. J. Chem. Eng. **45**, 66 (1967).

9. Borner, M.and Jacobasch, H.J., in "Symposium on Elektrokinetische Erscheinungen' Dresden 227 (1985).

XENON-NMR: A NEW PROBE FOR STUDYING STRUCTURAL

CHANGES OF POLYMERIC SURFACTANTS

Saeed Mohseni-Hosseini*

Department of Chemistry
University of Massachusetts
Amherst, Massachusetts   01002

ABSTRACT

In previous work we have shown the extreme sensitivity of the chemical shift of Xenon in solution which arises mostly from Van der Waals interactions. The sensitivity of the chemical shift to the local environment made possible the study of very low concentrations of aqueous solutions of association polymers (AP). AP was a long chain (7200 molecular weight) polyoxyethylene which was end-capped with saturated hydrocarbon C12 or C16. Brij-58 surfactant, a close model to AP, was also studied along with mixed polymer-sodium dodecy sulfate surfactant systems. The association site of Xenon is believed to be in a hydrocarbon-rich region. We also carried out investigations of solubilizate effect on association. Co(AcAc)$_3$ solubilizate was chosen since the association site was more hydrophilic in nature. Surface tension studies which were complimentary to the NMR work is reported as well.

INTRODUCTION

The chemical shift of an atom is sensitive to its environment, especially for the Xenon atom, which has NMR shifts of almost 200 ppm with respect to the isolated gaseous atom when Xenon is dissolved in some common solvents[1] and up

---

*Present Address: Olin Chemicals, 100 McKee Road, Rochester, New York   14611

Surface Phenomena and Latexes in Waterborne Coatings and Printing Technologies, Edited by M.K. Sharma, Plenum Press, New York, 1995

167

to 4000 ppm in some Xenon compounds. Xenon shifts are much larger than ^{13}C and ^{19}F shifts due to its highly polarizable electron cloud.

The NMR spectrum of Xenon in solution can give useful information about the Xenon atom[2-4]. Xenon is non-polar and chemically inert and, consequently, has minimal effect on its surroundings in non-aqueous media. Natural Xenon contains two magnetic isotopes, each in *ca.* 25% abundance, that have different nuclear properties. [129]Xe has a spin I = 1/2; it gives a sharp NMR line which allows accurate determinations of chemical shifts. This shift is very sensitive to the nature of the surrounding medium.

The medium shift of Xenon arises primarily from Van der Waals dispersive and repulsive interactions between the Xenon atom and its surroundings, while smaller contributions originate in the electric moment and magnetic anisotropy of the solvent[5-6].

[131]Xe has a spin I = 3/2; it has an electric quadrupole moment which causes short relaxation times and leads to broad NMR line. The relaxation rate is also sensitive to the environment and it is due to the quadrupole term, *i.e.*, the interaction of the nuclear electric quadrupole moment with electric field gradients generated by the surroundings[8-10].

The surface tension of aqueous solutions of anionic sodium dodecy sulfate (SDS)[7], non-ionic Brij 58 and AP 103-3 polymer was measured. In order to determine if the presence of a solubilizate has an effect on the CMC, cobalt acetyl acetonate (0.01 M Co(AcAc)$_3$) was added to the surfactant solutions.

EXPERIMENTAL

[131]Xe AND [129]Xe NMR

Both JEOL-90 MHz and XL-300 MHz NMR instruments have been used to measure the [129]Xe NMR chemical shift and [131]Xe relaxation rates in solutions of Brij 58 and the AP 103-3 polymer. Seven hours of accumulation are required to obtain a reasonable signal-to-noise ratio from a sample saturated with 16 atm. of Xenon using high pressure NMR tubes on the JEOL-90 Spectrometer. XL-300 NMR Spectrometer acquired only 2 hours of accumulation for an 8 atm. Xenon saturated solution.

It is observed that the raw [129]Xe chemical shift data vary from one spectrometer to the other for the sample. This is due to the effect of the bulk susceptibility of the sample and the use of an external reference (Xenon gas). The form of the correction differs for the two spectrometers and it has not been included here.

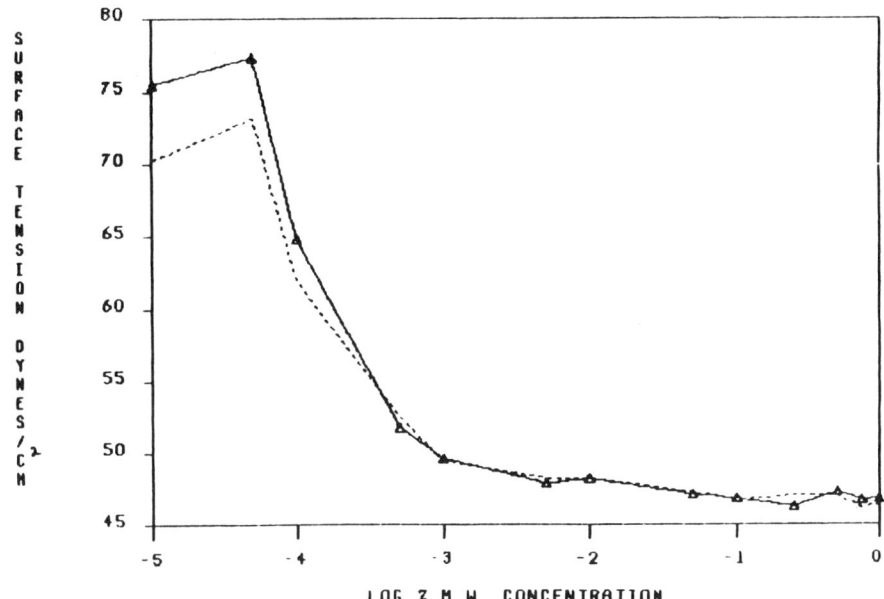

Figure 1. Surface Tension Versus Log of % M.W. Concentration for CMC Determination of Brij-58 Surfactant with (- - - - - - - -) and (————————) without Co(AcAc)$_3$ Solubilizate.

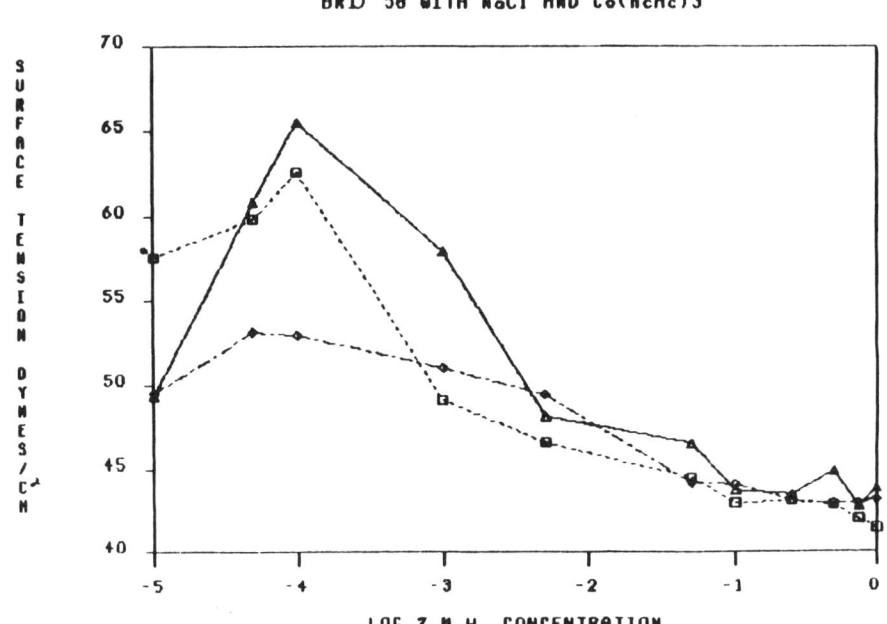

Figure 2. Surface Tension Versus Log of % M.W. Concentration for Brij-58 Surfactant with 0.01 M Co(AcAc)$_3$ at NaCl Concentrations of 0.5 M (Δ); 0.05 M (□) and 0.005 M (◇).

169

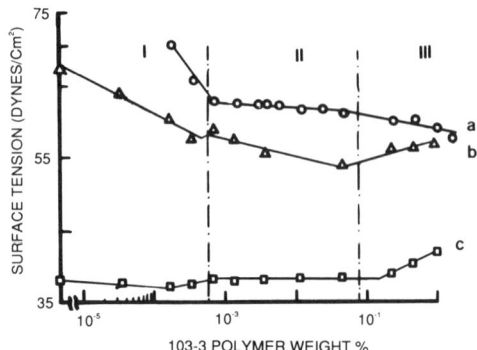

Figure 3. Surface Tension Versus 103-3 Polymer Weight % in the Absence (a) and in the Presence of 1 x 10^{-4} M SDS (b) and 5 x 10^{-2} M SDS.

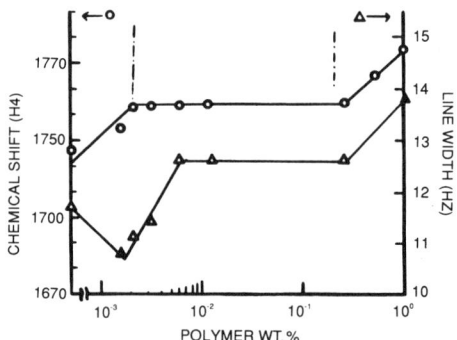

Figure 4. 103-3 Polymer Weight % Plotted Against ^{129}Xe Chemical Shift (O) and ^{131}Xe Line Width (Δ).

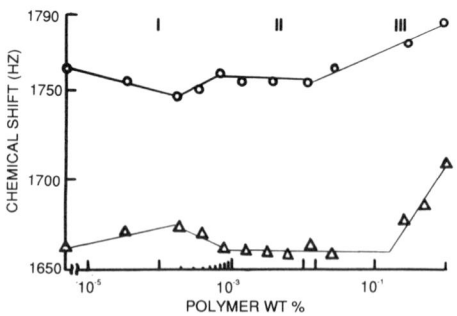

Figure 5. ^{129}Xe Chemical Shift Versus 103-3 Polymer Weight % with 1 x 10^{-4} M (O) and 5 x 10^{-2} M (Δ) of SDS Surfactant.

SURFACE TENSION

The Du Nouy ring tensiometer method was used; the measurements were carried out at room temperature and pressure. The ring tensiometer was calibrated each time using distilled water. The surface tension at each concentration of surfactant was measured at least two or three times, and the number reported is the *average* of those measurements.

REAGENTS

Brij 58 was supplied by Aldrich. SDS was purchased from Bio-Rad Laboratories (Richmond, CA). Co^{III} $(AcAc)_3$, tris acetylacetonate cobalt complex was obtained from two sources: one sample was synthesized by L. S. Frankel which contained some impurity, and the more purified sample was purchased from Harshaw Chemical Company (Cleveland, OH). The polymer was purchased from Union Carbide and it was called *"the associated polyethylene oxide 103-03 polymer."* Hereinafter, it appears as the 103-3 polymer.

RESULTS AND DISCUSSION

The surface tension measurements for Brij 58 model surfactant are shown in Figures 1-2. The CMC was determined to be around 10^{-3} Mol./Lit. which was not effected when 0.01 M of $Co(AcAc)_3$ salt was added. Lower concentrations below the CMC, however, were effected by $Co(AcAc)_3$ which lowered the surface tension in some cases as much as 5 dynes/cm^2 shown in Figure 1. When NaCl was added to the same Brij 58 solutions containing 0.01 M/Lit. $Co(AcAc)_3$, all concentrations especially below the CMC were effected. There seems to be an optimum NaCl concentration with the highest decreasing effect on the surface tension. As shown in Figure 2, the lowest NaCl concentration of 0.005 M had the highest effect on lowering the surface tension of Brij 58, indicating that very little ionic species are needed to enhance the micellar activities. Different behavior was observed for 103-3 polymer molecules when surface tension and Xenon NMR were performed. Surface tension measurements developed two breaks: one at lower 103-3 concentrations around 10^{-3} weight % and the other at about 10^{-1} weight % as shown in Figure 3a. This could indicate that these large size AP molecules may not necessarily form a regular micelle, instead they go through three different structural transition periods whether by self-association, dimer, trimer or number of molecules associating with each other. When 1 x 10^{-4} M (below CMC) sodium dodecy sulfate (SDS) surfactant was added to all 103-3 concentrations, surface tensions in general were lowered by 4-10 dynes/cm^2 and yet the two breaks could still be recognized as shown in Figure 3b. The higher SDS concentration of 5 x 10^{-2} M (above CMC), on the other hand, had a major effect on lowering their surface tensions by almost 25 dynes/cm^2. The higher SDS concentration also altered the two breaks completely. The first break at low 103-3 concentrations almost disappeared while the second break at high 103-3 concentrations changed configuration from negative to positive slope as shown in Figure 3c. It would be difficult to interpret these behaviors; the high SDS concentration seems to form mixed micelle with all polymer concentrations and surface tension values are very close to SDS micelle itself[7]. The positive slope, however, could indicate change in micellar configuration from one form to another.

It is interesting that ^{129}Xe and ^{131}Xe NMR studies of 103-3 polymer bulk solvents produced almost the same supporting results. Like surface tension, both ^{129}Xe chemical shift and ^{131}Xe line width plotted against 103-3 polymer concentrations generated two distinct breaks shown in Figure 4. Breaks developed by NMR are indicating that bulk solutions are verifying the polymer structural (self, dimer, trimer association) transitions determined earlier by surface activities. Two different SDS concentrations, one below (1 x 10^{-4} M) and the other above (5 x 10^{-2} M) its CMC, were added to various 103-3 polymer concentrations and ^{129}Xe NMR chemical shift studies were conducted as shown in Figure 5. These NMR studies of mixed polymer-surfactant solutions once again supported surface activity behaviors. ^{129}Xe NMR chemical shift was hardly effected for SDS addition below its CMC (Figure 5) while it was drastically influenced by SDS addition above its CMC (Figure 5), just as shown by surface activitiy measurements. It appears that high SDS concentration added to the polymer could over dominate the polymer activities.

Xenon is non-polar inert gas which preferentially dissolves in non-polar environment[8-10] or the non-polar region of micelle and verifies micellar formation in bulk solution[7]. This makes Xenon a high candidate for CMC determination based on bulk solution which could be more reliable than CMC determined by surface tension since it eliminates aerosol effect.

In conclusion, Xenon NMR is an excellent probe to study bulk solvent behavior of surfactants, polymeric surfactants or polymer-surfactant mixtures and results normally support surface tension studies.

REFERENCES

1. Schrobilgen, G.J., NMR and the Periodic Table, Eds., R.K. Harris, and B. E. Mann, 439 (1978).

2. Miller, K.W., Reo, N.V., Schoot Uiterkamp, A.J.M., Stengle, D.P., Stengle, T.R. and K. L. Williamson, K.L., Proc. Natl. Acad. Sci., USA, **78**, 4946, (1981).

3. Ripmeester, J.A. and Davidson, D.W., J. Mol. Structure., **75**, 67, (1981).

4. Tilton, R.F., Jr. and Kuntz, I.D., Jr., Biochemistry, **21**, 6850, (1982).

5. Rummens, F.H.R., "Van der Waals Forces and Shielding Effects. NMR Principles and Progress." Vol. **10**, New York. Springer Verlag, (1975).

6. Stengle, T.R., Reo, N.V. and Williamson, K.L., J. Phys. Chem., **85**, 3772, (1981).

7. Mohseni-Hosseini, S., Ph.D. Thesis, University of Massachusetts, Amherst, MA., (1985).

8. Frankel, L.S., Langford, C.H. and Stengle, T.R., J. Phys. Chem., **74**, 1376, (1970).

9. Stengle, T.R., Hosseini, S.M., Basiri, G.H. and Williamson, K.L., J. Solution. Chem., **13**, 779, (1984).

10. Stengle, T.R., Hosseini, S.M. and Williamson, K.L., J. Solution. Chem., **15**, 777, (1986).

THE HIGH SHEAR RHEOLOGICAL PROPERTIES

OF DISPERSIONS

V. Ribitsch and J. Pfragner

Department of Physical Chemistry
University Graz
Heinrichstr. 28
A-8010 Graz
Austria

ABSTRACT

Extremely high shear rates are applied to pigment dispersions during various scale-up and manufactring processes in several industries. As reported in this paper, one has to know the rheological properties of dispersions under high shear, which can not be estimated but have to be measured under similar conditions. Carbonate and clay slurries exhibit a maximum in viscosity with increasing shear rate, responsible for failures in coating processes. Several additives under investigation shift the maximum towards higher shear rates and reduce its height. Water-soluble polymers might even level it off, depending on their molecular weight and molecular weight distribution. It is concluded that the polymers stabilize the liquid layer around the particles. Hence, shear forces may disaggregate flocs, but will not increase the interparticle repulsion.

INTRODUCTION

China Clay and $CaCO_3$ slurries containing surfactants, latices, different water soluble polymers and other additives are handled at fairly high concentrations (solid

Surface Phenomena and Latexes in Waterborne Coatings and Printing Technologies, Edited by M.K. Sharma, Plenum Press, New York, 1995

175

contents about 55% to 70% by weight) as paper coatings. Their rheological properties are crucial for the runability on high speed paper coating machines where shear rates between 10^5 to 10^6 1/s are common[1-3]. Dilatancy phenomena frequently occurring with highly concentrated suspensions at high shear rates cause severe problems in production. A proper coating formula prevents these problems. It is therefore, necessary to understand the influence of additives on the rheological behavior of the suspensions especially at high shear rates. This applies not only for paper coating colors, but also for other pigment slurries at similar production processes. In all these cases, it is impossible to predict the high-shear processing properties of these pigment dispersions from rheological data collected at low to medium shear rates.

EXPERIMENTAL

The pigment dispersions in water were prepared using a high speed stirrer (700 RPM). All dispersions were, unless otherwise mentioned, stabilized with 0.3% anionic surfactant (polysalt). Each sample was stirred for 2 hours at constant speed (500 RPM) before the rheological experiments were performed.

The rheological experiments were performed with a high shear capillary viscometer (HVA-6 A.Paar KG, Graz, Austria)[4]. A schematic diagram of the instrument is shown in Figure 1.

This instrument uses nitrogen as driving pressure (1 to 160 bar). The sample is palced in a pressure chamber and separated from the surrounding thermostat fluid by a rubber bag. The driving pressure forces the sample through an attached capillary, the pressure at the capillary entrance and the flow rate are recorded.

RESULTS

The experimental data obtained are described as follows:

VISCOSITY BEHAVIOR OF PIGMENT DISPERSION WITHOUT ADDITIVES

The viscosity as a function of shear rate depends, beside the concentration very much, on the nature of the pigment, the particle size and the particle size distribution.

CaCO$_3$ DISPERSIONS: Suspensions of CaCO$_3$ pigments (Hydrocarb 60, particle size 60% smaller then 2 μm) don't show rheological abnormalities, they are insignificantly shear thinning at shear rates below 1000 1/s, exhibit constant viscosity between 1000 and 10.000 1/s and only a slight viscosity increase at high shear rates if the solid content exceeds 65%. Below this concentration almost newtonian behavior is observed.

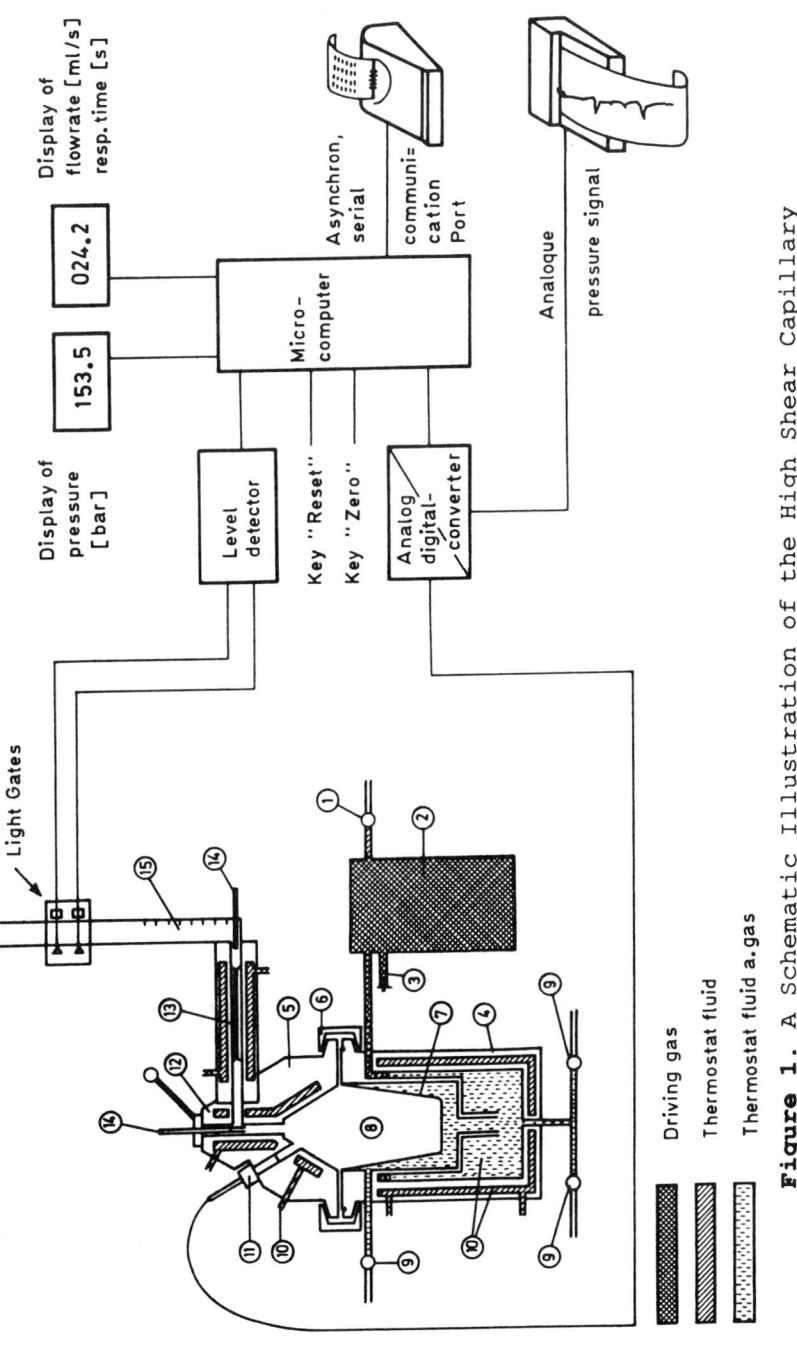

Figure 1. A Schematic Illustration of the High Shear Capillary Viscometer HVA 6. (1) Pressurizing System (Pressure Regulating Valve, Nitrogen Tank), (2) Pressure Detection (Pressure Transducer, p min/p max Discriminator), (3) Pressure Chamber, (4) Thermostat Unit, (5) Measuring Device (Capillaries 1 = 0.1 to 100 mm, Slits), (6) Flow Measuring Unit, (7) Data Collection and Evaluating Unit.

CaCO$_3$ slurries (Hydrocarb 90, particles size 90% smaller then 2 μm) are somewhat different (Figure 2). They exhibit a viscosity maximum at concentrations above 65% and shear rates higher then 104 1/s. This viscosity maximum increases with increasing concentration, for example the viscosity of a 75% slurry at a shear rate of 70.000 1/s is three-fold of that at 4000 1/s.

CLAY DISPERSIONS: Suspensions of China Clay (Figure 3) show a rheological behavior strongly depending on concentration and shear rate. At low solid contents below 50% almost newtonian flow curves are obtained.

Increasing solid contents change the viscosity function considerably. The low shear behavior is dominated by the increasing yield stress and exhibits therefore high apparent viscosity, rapidly decreasing with increasing shear rate. At the high shear region an other effect onsets and becomes remarkable at concentrations above 50%.

At shear rates above 10^4 1/s, a viscosity increase and a viscosity maximum around 10^5 1/s is observed. This maximum is shifted towards lower shear rates around 10^4 1/s with increasing solid contents, and the maximum viscosity increases gradually.

The possible reasons of the viscosity increase can be discussed as a change in particle structure and a change in particles effective volume. Indeed, those rheological data maintain further information about the hydrodynamically effective volume and structure of the particles. One possible way to obtain information about the interaction between the solid and the solvent is the procedure described by Mooney[5].

$$\frac{1}{\ln \eta_r} = - \frac{1}{[\eta]} \cdot \frac{1}{\Phi_i} + \frac{1}{[\eta]} \cdot \frac{1}{\Phi}$$

$[\eta]$ = Volume limiting viscosity number, a unit for viscometric efficient particle - the equivalent sphere (Including particles, excluded volume adsorbed and immobilized dispergant)

Φ_i = Immobilization concentration, the concentration at which the maximum package of particles is reached and where the viscosity is infinity - no flow appears.

The relative viscosity values at different concentrations and constant shear rates are correlated to the volume fraction of the solid content. Following Mooney the results should be straight lines at any given shear rate. From the slope one can calculate the limiting viscosity number, which describes the hydrodynamical volume of the particle. From the intercept one can obtain the immobilization concentration, which describes the state of maximum package of the dispersed particles.

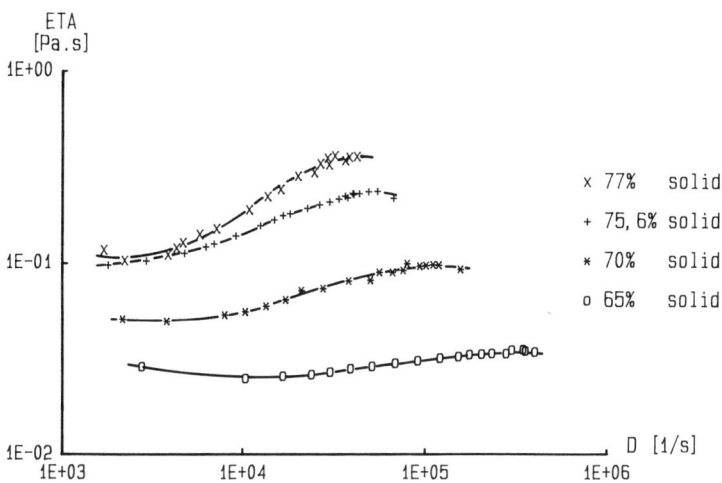

Figure 2. Viscosity Curves of CaCO3 (90% of the Particles < 2 μm).

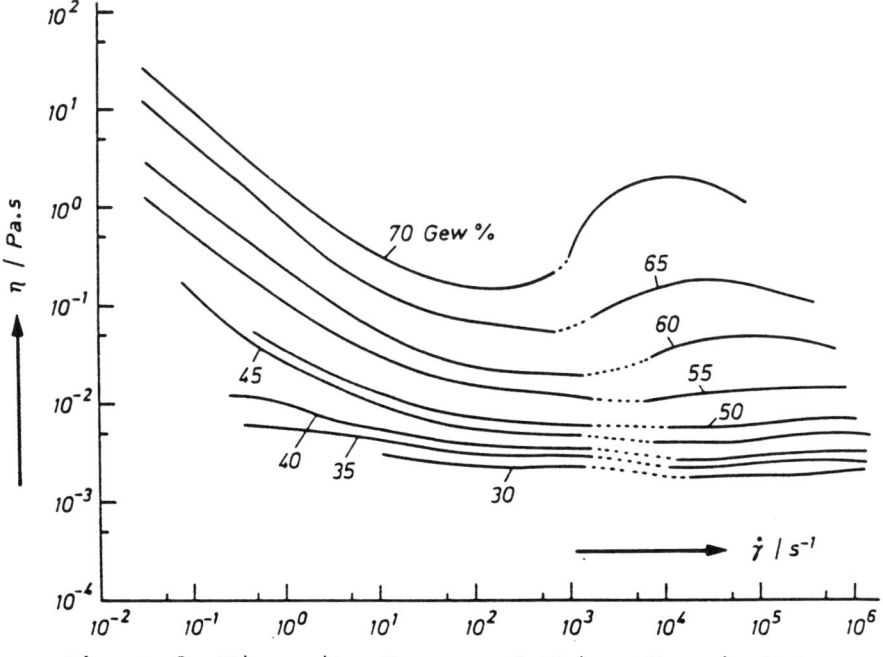

Figure 3. Viscosity Curves of China Clay in Water.

The Mooney plot in Figure 4 shows two straight lines describing the typical flow situation at the low shear rate and high shear rate region, respectively. Slope and intercept of these graphs indicate, that the volume limiting viscosity number $[\eta]$ decreases with increasing shear rate; the following values were obtained:

low shear rates: $[\eta]$ from 9 to 5 (g/g)

high shear rates: $[\eta]$ from 3.95 to 3.5 (g/g)

The immobilization concentration also decreases from

low shear rates: Φ_i = 0.85

high shear rates: Φ_i = 0.59

These rheological properties may be explained in the following way[6]: Suspensions of China Clay exhibit at low shear rates weakly agglomerated (flocculated) particles causing the high yield stress. The adhering dispersant contains the counter ions balancing the particles charge. Increasing solid content causes closer packing of the agglomerates and therefore, demands higher shear stresses for flow onset.

At high shear rates the aggregates are mainly destroyed leading to a suspension of a rather high number of small particles as indicated by the reduced limiting viscosity number. Moreover, the particles lose a part of the adhering dispersant including a part of the counter ions. This causes strong repulsive forces and an increase of viscosity.

In this way two different states of the same system are proposed[7], one with agglomerates and balanced particle charge, and the other one of particles which considerably lost the contact to their counter ions. Both may be described by continuously decreasing viscosity functions. One state exists under low shear rates, the other one with great viscosity contribution at high shear rates. The transition from one to the other causes the maximum in viscosity and may be called a transient dilatancy. This intermediate increase of the viscosity is the reason of instabilities and the well known production problems.

A very similar flow behavior is found with clays from different sources (Figure 5). The US Clay No. 1. shows almost the same rheological properties as Amazon. The clay KKN which causes even more problems during coating shows a more pronounced shear thickening then the Amazon clay. Immobilization is reached at a concentration of 68 % and shear rates above 10^5 1/s.

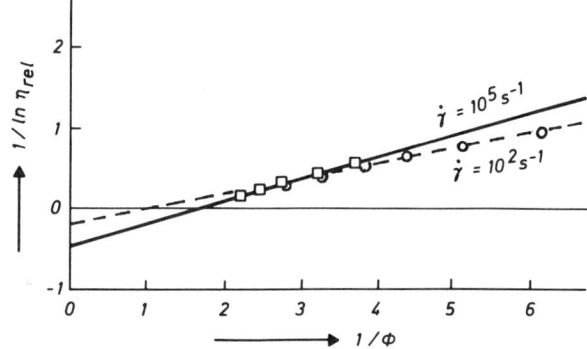

Figure 4. Mooney Plot - Amazone in Water at Different
Shear Rates.

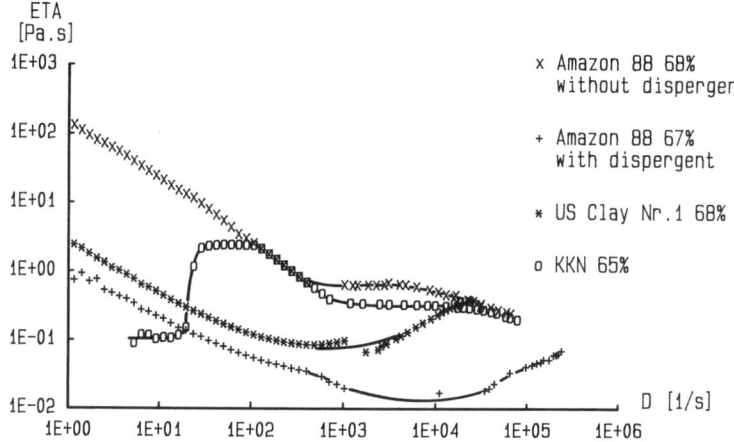

Figure 5. Comparison of the Flow Properties of Several
Clay Dispersions.

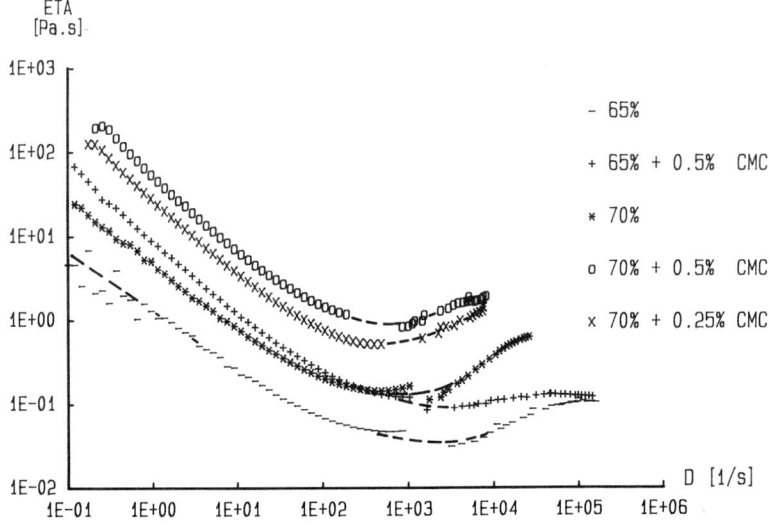

Figure 6. Rheological Properties of Clay Dispersions as
a Function of CMC Concentration.

INFLUENCE OF POLYMER ADDITIVES ON THE RHEOLOGICAL PROPERTIES OF MINERAL SUSPENSIONS: Watersoluble polymers (carboxymethylcellulose, CMC) influence the rheological properties of pigment dispersion depending on the molecular weight (M_W) and the molecular weight distribution (MWD) of the polymer.

The addition of CMC to clay dispersions increases the viscosity and reduces the dilatancy depending on the applied concentration (Figure 6). 0.25% CMC cause a decrease of the high shear rate viscosity (reduction of dilatancy) and shift the viscosity maximum towards high shear rates. The low shear viscosity is of coarse also increased. The addition of 0.5% CMC causes the dilatancy phenomenon to vanish, resulting in a suspension showing shear thinning at low to moderate and an almost constant viscosity at shear rates above 10^4 1/s.

This is also observed with different clay types and $CaCO_3$ (Figure 7). KKN needs more CMC then Amazon to reduce the dilatancy and shift it towards higher shear rates, also it never vanishes.

The molecular weight and molecular weight distribution also influence the rheological properties: Figure 8 shows the flow curves of 70% Amazon with 0.5 % CMC of different qualities.

The addition of a medium molecular weight CMC with narrow molecular weight distribution does not alter the flow curve but increasing the viscosity level. The addition of CMC with broad molecular weight distribution causes a reduction of the dilatancy and a shift towards higher shear rates. A wide range of constant viscosity at 10^3 to $5*10^4$ 1/s is observed. A similar effect is found with CMC of high molecular weight and broad molecular weight distribution.

From the rheological data one can calculate a 'slip - coefficient' describing the strength of the particle - dispersant interaction. It is calculated according equation 2[8]. Figure 9 presents this data for the same CMC qualities as described above.

$$\frac{Q}{r^3 \cdot \pi} = \frac{\tau}{4 \cdot \eta} + \frac{1}{r} \cdot v_G$$

where, v_G = slipping velocity

The clay dispersion without polymer shows the highest sliding coefficient, indicating the weakest interaction between the pigment and the dispersant. The CMC sample with medium molecular weight and narrow molecular weight distribution reduces the slip coefficient but it is still the highest one of all CMC samples. The sample with broad molecular weight distribution - containing a bigger amount of high molecular species - has a smaller slip coefficient,

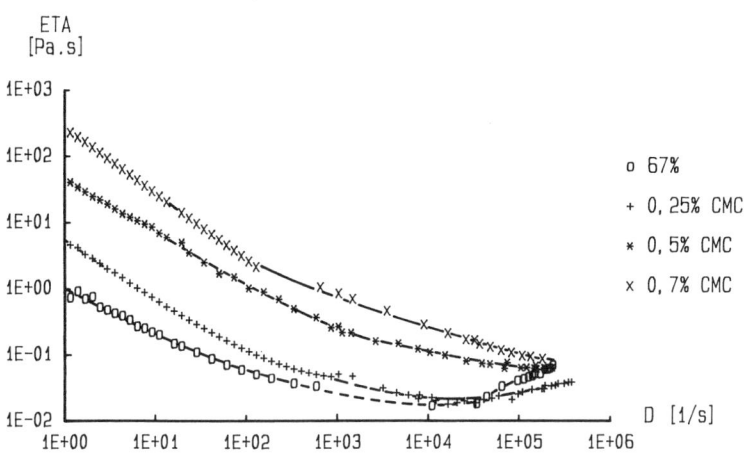

Figure 7. Flow Properties of CaCO3 Slurries as a Function of Solid Content and CMC Concentration.

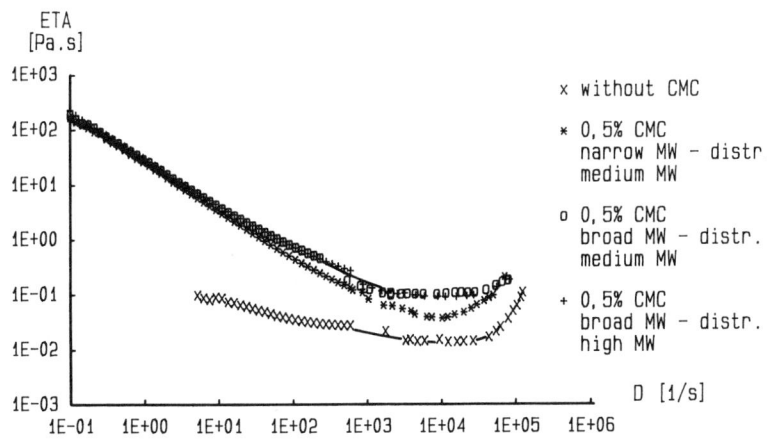

Figure 8. Flow Curves of 70% Amazon With 0.5 % CMC of Different Molecular Weight.

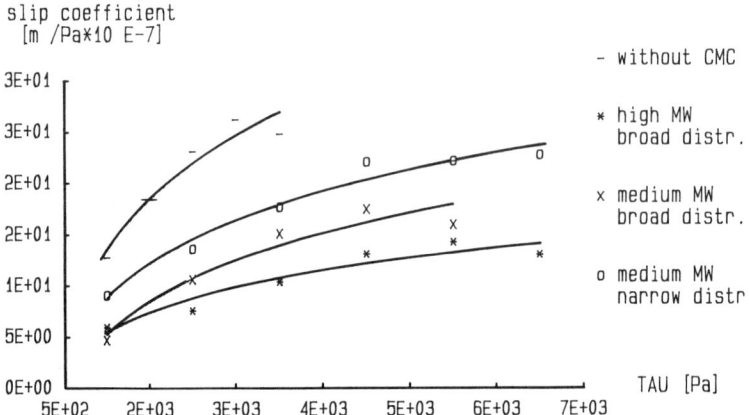

slip coefficient
[m /Pa*10 E-7]

- without CMC

* high MW
  broad distr.

x medium MW
  broad distr.

o medium MW
  narrow distr.

TAU [Pa]

Figure 9. Slip Coefficient as a Function of Shear Stress for Clay Slurries With Different CMC Types.

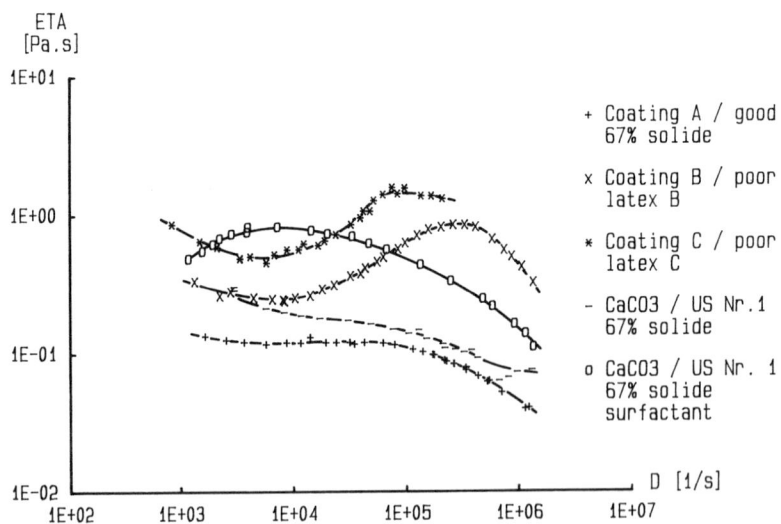

ETA
[Pa.s]

+ Coating A / good
  67% solide

x Coating B / poor
  latex B

* Coating C / poor
  latex C

- CaCO3 / US Nr.1
  67% solide

o CaCO3 / US Nr. 1
  67% solide
  surfactant

D [1/s]

Figure 10: Rheological Properties of Different Paper Coating Colors.

184

and the high molecular weight CMC decreases it further. This data indicate an increase of pigment - dispersant interaction with increasing amount of high molecular weight CMC.

HIGH SHEAR VISCOSITY OF PAPER COATING COLORS: Paper coatings are widely used at high shear rates. Their rheological properties are responsible for the runability in coating aggregates. This rheological properties are influenced by the type of pigment, the applied polymer binder and the added latex. Some examples are shown in Figure 10.

Coating A and the $CaCO_3$/US Clay No. 1 slurry with different polymer additives don't exhibit rheological instabilities and perform satisfactory in high shear coating machines. Paper coatings with good processing characteristics show this type of viscosity functions with viscosity values at high shear rates between 0.05 to 0.25 Pa.s. The more viscous one of the $CaCO_3$/Clay slurries can be handled under processing conditions.

Coating color B and C cause poor coating quality. Both colours show an increase of viscosity of approximately the three-fold in the shear rate range of high speed machine conditions. In general, it is experienced that all coatings show dilatancy phenomena, which cause problems during processing.

DISCUSSION

The flow properties of pigment slurries are mainly determined by the particle-particle and particle-dispersant interaction. These can be modified by means of surfactants and polymer additives. The systems are very sensitive to shear forces. Moderate shear rates, as coming along with maintaining the stock slurries, may deflocculate weakly aggregated particles. Strong forces due to the coating process may alter the surface and neighbourhood of the

particles and influence in this way the interaction between particles and dispersant. As the liquid close to the particle surface is sheared off, charges are seperated. This gains interparticle repulsion and enhances the viscosity of the slurries. The results presented show the viscosity increase under high shear forces and the possibility of additives in order to control the crucial viscosity maximum.

The addition of water-soluble polymers enforces the immobilized liquid with respect to the particle and reduces in this way the shear induced viscosity increase. This effect depends on the polymer molecular weight. The higher the amount of long chain molecules, the more pronounced is the depression of the viscosity increase. This is reasonable because the higher the molecular weight the stronger is the polymer adsorption on the pigment particles. The adsorbed hydrophilic polymers stabilize a layer of immobilized water and counterions around the particle.

The better understanding of the colloidal mechanism under high shear conditions may help to optimize the coating formulas i.e. to increase the solid content and to improve the processability of pigment dispersions.

REFERENCES

1. Ribitsch, V., Proceedings of the 35th Rheology Symposium of the Japanese Rheological Society, 57-61, (1987).

2. Laun, H.M., Hirsch, G.L., Rheol. Acta **28**, 267-272, (1987).

3. Eklund, D., Strand, M.,Wochenblatt für Papierfabrikation **15**, 577, (1980).

4. Ribitsch, V., Ch. Jorde, J. Schurz, GIT **26(12)**, 1112-1116, (1982).

5. Mooney, M., Trans. Soc. Rheol., **2**, 210-216, (1931); J Colloid Sci., **6**, 162 (1951).

6. Pfragner, J., Progr. in Colloid & Polymer Sci., **77**, 177-183, (1988).

7. Schurz, J., Wochenblatt für Papierfabrikation, **8**, 275-279, (1984).

8. Jastrzebinski, J., EC Fundamentals, **6(3)**, 445-454, (1967).

THREE-DIMENSIONAL CHARACTERIZATION OF ACTIVE

SURFACTANT

Victor P. Janule

SensaDyne Instrument Division
Chem-Dyne Research Corporation
P.O. Box 30430
Mesa,
Arizona 85275-0430

ABSTRACT

Active surfactant are known to have dynamic characteristics when surface tension is determined as a function of interface development time. Higher surface tensions result as interface development times are reduced. This allows the characterization of a fluid in two dimensions, surface tension versus interface development time, which includes both the static (or equilibrium) and dynamic zones. When a fluid's surface tension is further measured as a function of surfactant concentration, then the action of the surfactant can be characterized in three dimensions: concentration, surface tension, and interface development time. The method presented can be applied readily to active surfactant added to coatings, aqueous solutions, fountain solutions/ink emulsions, and other chemicals or formulations for which surface tension can be measured using the modified maximum bubble pressure method. In the example presented, the SensaDyne® Tensiometer was used with several auxiliary software programs, one for accurate determination of interface development times, and a second program for three-dimensional graphing of the data.

Surface Phenomena and Latexes in Waterborne Coatings and Printing Technologies, Edited by M.K. Sharma, Plenum Press, New York, 1995

187

INTRODUCTION

The SensaDyne® Tensiometer uses a patented technology that is a refinement of the maximum bubble pressure method. This method was first suggested by Simon in 1851 and later developed by Jaeger in 1917. The first viable commercial instrument was introduced in 1982, and a subsequent design interfaced to the personal computer several years later, allowing us now to use software tools that make three dimensional studies relatively straight forward.

This modified maximum bubble pressure method is illustrated in Figure 1, showing a descriptive diagram of two capillaries of different radii immersed beneath a fluid surface. Process gas, bubbled through these tubes, produces a differential pressure signal, the value of which is used to calculate fluid surface tension. The resulting differential pressure equation includes effects of fluid head (h), density of the liquid (ρ), and gravitational constant (g). These are effectively cancelled if the orifices are oriented to allow the bubbles to release at the same height. This normal probe arrangement can be modified to use inverted probes to mitigate problems of bubble distortion that can result in viscous coatings or fluids with a high solids content. These inverted probes allow unobstructed upward release of the bubbles in the fluid.

The rate of gas flow to the orifices is a user-controlled feature of the SensaDyne® Tensiometer, allowing the bubble rate to be varied, and consequently the interface development time. The amount of time it takes the bubble to form and release from the orifice is also the amount of time available for the surfactant to migrate to the interface where it can lower the surface tension.

By measuring surface tension at different bubble rates in a fluid or coating that has an active surfactant, a dynamic surface tension graph can be plotted similar to the one shown in Figure 2. In this example, the dynamic curves of two different fluids are plotted. The curve results when the values obtained at different bubble rates are connected. The transition from the dynamic to the static zone occurs where the curves level off horizontally.

The instrument operates by generating a sawtooth wave from the electronic output of a differential pressure transducer, as illustrated in Figure 3. This is an accurate representation of the pressure waveform that occurs at the small orifice. The electronics track each bubble as it forms until a peak is reached [the maximum differential bubble pressure]. This peak value is captured and then updated with each subsequent peak value. The resultant steady state output is an effective "moving average" of discrete surface tension measurements that accurately tracks the surface tension of the test fluid or coating.

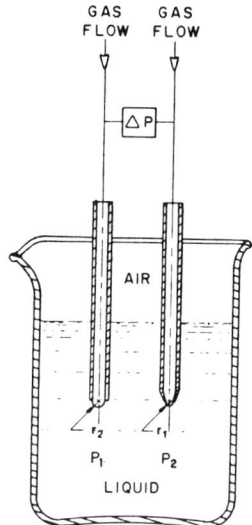

Figure 1. A Schematic Diagram of Measuring Dynamic Surface Tension by Bubble Method.

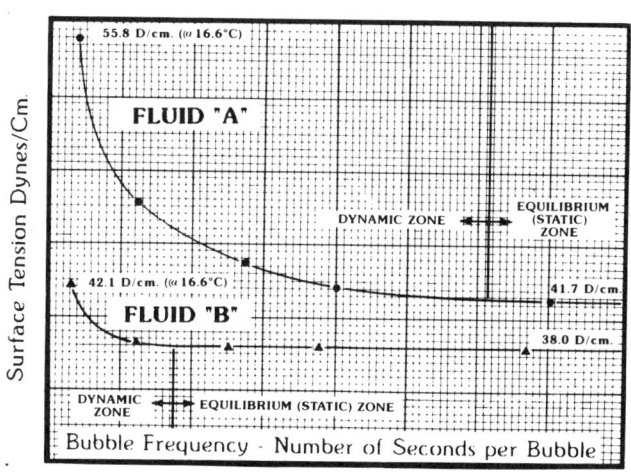

Figure 2. Dynamic Surface Tension Versus Bubble Frequency (Number of Seconds/Bubble).

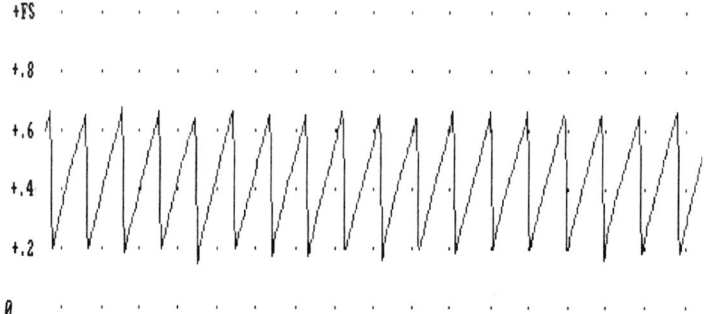

Figure 3. Differential Pressure Transducer Output.

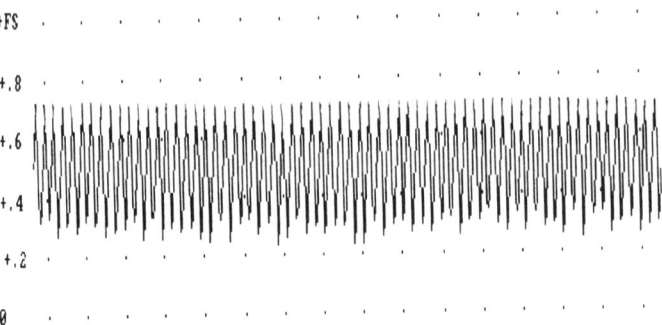

Figure 4. Transducer Output Versus Fast Bubble Rate.

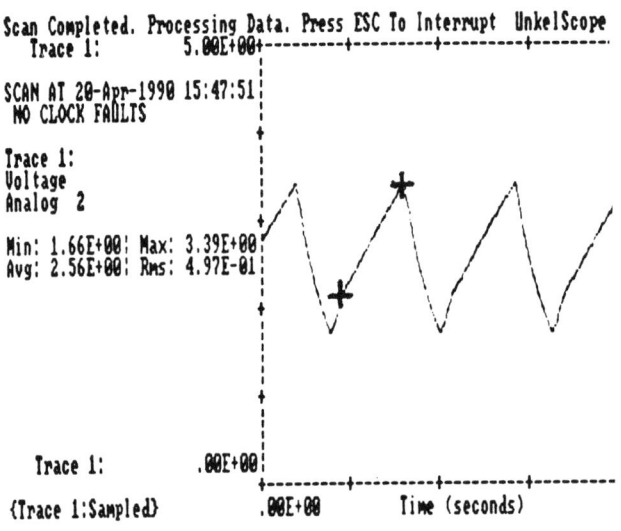

Figure 5. Oscilloscope Program: Valid Bubble Rate Setting.

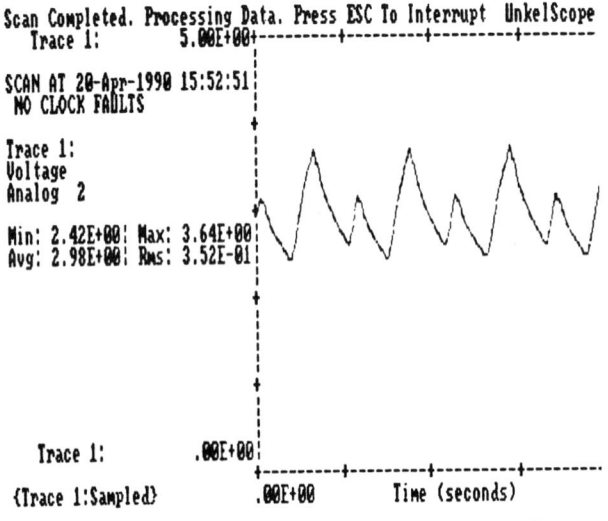

Figure 6. Oscilloscope Program: Valid Bubble Rate Setting.

This technology allows the instrument to be used for intermittent or continuous measurements at any point on the dynamic curve where an active surfactant is present. The continuous measurement capability enables process control using an enhanced version of the same software program [UnkelScope®] that is used for interface development time verification.

At slow bubble rates, the interface development time is approximately equal to the bubble rate [peak to peak value] that is displayed on the computer screen. If desired, this can also be stored on disk in a data file along with surface tension values and fluid temperature. As the bubble rate is increased, as illustrated in Figure 4, it becomes increasingly difficult to verify the bubble formation time from the computer software's normal strip chart program option.

Verification becomes increasingly important as the bubble rate increases because the interface development time becomes a decreasing percentage of the peak to peak time [bubble rate]. A second problem occurs because a specific-sized orifice will only generate a limited number of discrete bubbles per second as the flow rate is increased before the bubbles are forced out so fast that they begin to link together. For example, a 0.5 MM. orifice can support approximately ten to twelve discrete bubbles per second while a 0.25 MM. orifice can go as high as twenty to twenty-five bubbles per second.

To address these problems, a supplemental software program is used to turn the computer into an oscilloscope, with the screen allowing a display such as is illustrated in Figure 5. This display covers a one second time frame, showing a frequency of four bubbles per second. An auxiliary feature of this software allows any display to be captured in memory. It can be brought back to the screen and then a pair of cursors can be moved along the waveform. One is positioned at the start of bubble formation and the other at the peak value (as illustrated). The time between cursors is automatically calculated and displayed in the left side data column (not shown) allowing the accurate determination of bubble formation time and consequently the maximum allowable surfactant migration time.

Figure 6 illustrates a condition that approaches an oscillating jet; the flow is too great for the orifice size and bubbles have started to link together. In this particular example each large peak represents a single discrete bubble. Each small peak represents a pair of bubbles that are linked together.

The progression from Figures 5 to 6 would be an increasing number of large peaks until a single small peak would result as the first pair of bubbles link together. As the flow rate increases, the number of small peaks would increase until eventually the large peaks disappear and no single bubbles emerge from the orifice. Because the instrument cannot distinguish between a valid peak (maximum bubble pressure) and a small (false) peak, it will average the waveform, causing an invalid, and often fluctuating, surface tension output reading.

At fast bubble rates, it is important to set the instrument up to generate valid data, by generating a series of individual (non-linked) bubbles from the worst case fluid to be used. Otherwise there is a risk that the instrument will be calibrated and used, and the resultant data may not be valid. The worst case is the waveform and bubble rate that results in the lowest surface tension fluid to be measured, either for testing or calibration. Typically, the worst case is when alcohol is used as the lower surface tension calibration standard.

If alcohol is the only low surface tension calibration fluid available, this problem can be mitigated by calibrating the instrument with water and alcohol, formulating and accurately measuring a "new" water/alcohol mixture (for example, in the 40 to 50 dynes/cm. range), and using this mixture as the "new" low surface tension standard. This moves the worst case lower limit from the 20 dynes/cm. range into the 40 to 50 dynes/cm. range.

EXPERIMENTAL

MATERIALS

Regain NF is a heavy duty liquid stripper/degreaser that exhibits known dynamic characteristics due to an active surfactant. It has been tested at various times to generate dynamic curves. Solubility in water is complete, and is stable under normal handling conditions. It's overall characteristics are representative of water-based coatings, inks, fountain solutions, and most fluids that contain active surfactant. The dynamic curves for Regain NF are similar to the examples shown in Figure 2, for Fluids A and B. Regain NF is composed of the chemicals listed in Table I.

Table I. Composition of Regain NF

* Water, Zeolite Softened

* Sodium Silicate

* Monoethanolamine

* Butoxyethanol

* Polyethoxylated Alcohol

* Tetrasodium EDTA

Table II. Experimental Data in ASCII Format for Run (R-1.TXT).

DATE	TIME	#	TEMP.	B/S	S.T.	COMMENTS
06-25-1992	10:15:33	A	22.3	0.17	55.7	0.2 ml Concentration
06-25-1992	10:25:44	B	22.3	0.19	51.3	0.4 ml Concentration
06-25-1992	10:30:54	C	22.3	0.20	48.8	0.6 ml Concentration
06-25-1992	10:34:05	D	22.3	0.22	47.7	0.8 ml Concentration
06-25-1992	10:40:41	E	22.3	0.23	46.5	1.0 ml Concentration
06-25-1992	10:45:00	F	22.4	0.25	45.8	1.2 ml Concentration
06-25-1992	10:48:13	G	22.5	0.26	45.5	1.4 ml Concentration
06-25-1992	10:50:41	H	22.5	0.27	45.3	1.6 ml Concentration
06-25-1992	10:54:04	I	22.6	0.27	44.5	1.8 ml Concentration
06-25-1992	10:58:43	J	22.6	0.29	44.4	2.0 ml Concentration
06-25-1992	11:04:07	K	22.7	0.30	44.3	2.2 ml Concentration
06-25-1992	11:10:29	L	22.7	0.31	44.1	2.4 ml Concentration
06-25-1992	11:14:13	M	22.7	0.33	43.8	2.6 ml Concentration

Table III. Experimental Data in ASCII Format for Run (R-2.TXT).

DATE	TIME	#	TEMP.	B/S	S.T.	COMMENTS
06-25-1992	12:08:07	A	23.1	0.38	58.4	0.2 ml Concentration
06-25-1992	12:14:00	B	23.1	0.42	54.3	0.4 ml Concentration
06-25-1992	12:16:52	C	23.1	0.45	51.5	0.6 ml Concentration
06-25-1992	12:21:19	D	23.1	0.47	49.8	0.8 ml Concentration
06-25-1992	12:26:04	E	23.2	0.48	48.5	1.0 ml Concentration
06-25-1992	12:30:03	F	23.2	0.49	47.5	1.2 ml Concentration
06-25-1992	12:33:12	G	23.2	0.50	47.1	1.4 ml Concentration
06-25-1992	12:36:44	H	23.3	0.51	46.8	1.6 ml Concentration
06-25-1992	12:43:09	I	23.3	0.52	46.2	1.8 ml Concentration
06-25-1992	12:44:55	J	23.3	0.52	45.5	2.0 ml Concentration
06-25-1992	12:46:41	K	23.4	0.53	45.2	2.2 ml Concentration
06-25-1992	12:50:55	L	23.4	0.54	45.0	2.4 ml Concentration
06-25-1992	12:55:23	M	23.4	0.55	44.7	2.6 ml Concentration

Table IV. Experimental Data in ASCII Format for Run (R-3.TXT).

DATE	TIME	#	TEMP.	B/S	S.T.	COMMENTS
06-25-1992	13:54:46	A	23.4	0.98	63.0	0.2 ml Concentration
06-25-1992	13:57:01	B	23.4	1.07	59.0	0.4 ml Concentration
06-25-1992	14:00:27	C	23.5	1.14	55.9	0.6 ml Concentration
06-25-1992	14:03:11	D	23.5	1.20	53.8	0.8 ml Concentration
06-25-1992	14:05:58	E	23.5	1.24	52.1	1.0 ml Concentration
06-25-1992	14:09:48	F	23.5	1.28	50.7	1.2 ml Concentration
06-25-1992	14:13:14	G	23.6	1.31	49.8	1.4 ml Concentration
06-25-1992	14:16:49	H	23.6	1.33	48.8	1.6 ml Concentration
06-25-1992	14:19:32	I	23.6	1.35	47.7	1.8 ml Concentration
06-25-1992	14:21:19	J	23.6	1.38	47.5	2.0 ml Concentration
06-25-1992	14:25:48	K	23.7	1.39	47.2	2.2 ml Concentration
06-25-1992	14:28:36	L	23.7	1.40	46.8	2.4 ml Concentration
06-25-1992	14:30:38	M	23.7	1.43	46.3	2.6 ml Concentration

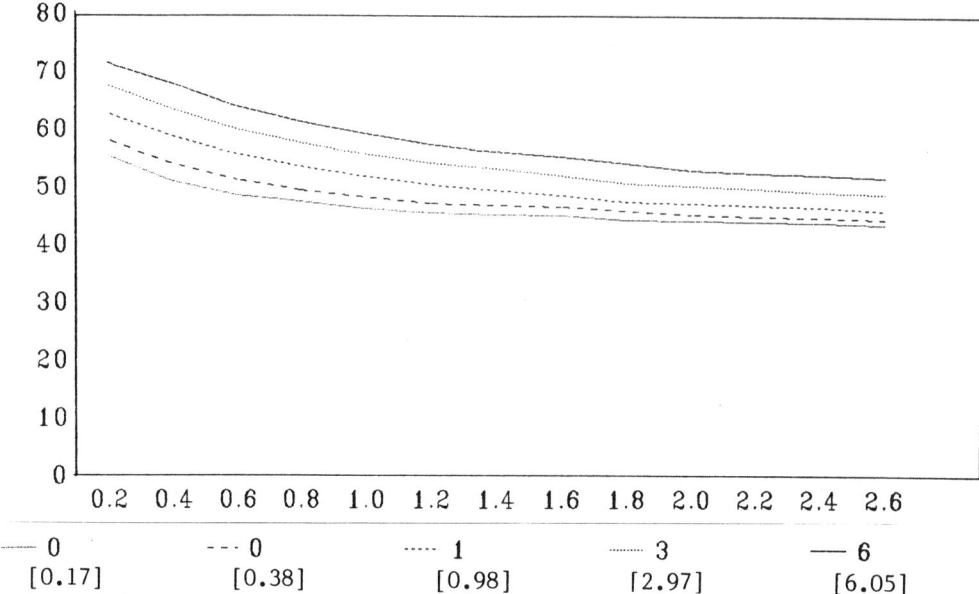

Figure 7. Surface Tension as a Function of Concentration (Two Dimensional Test Data Plots at Five Bubble Settings).

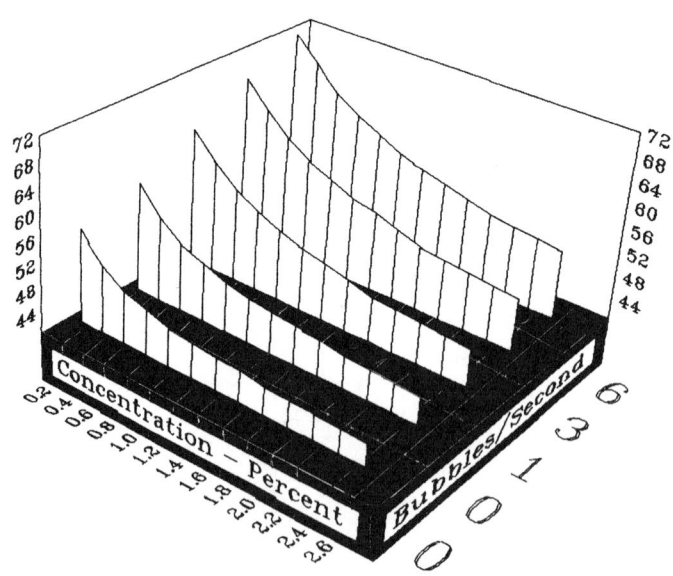

Figure 8. Surface Tension as a Function of Concentration (Two Dimensional Test Data Plots at Five Bubble Settings).

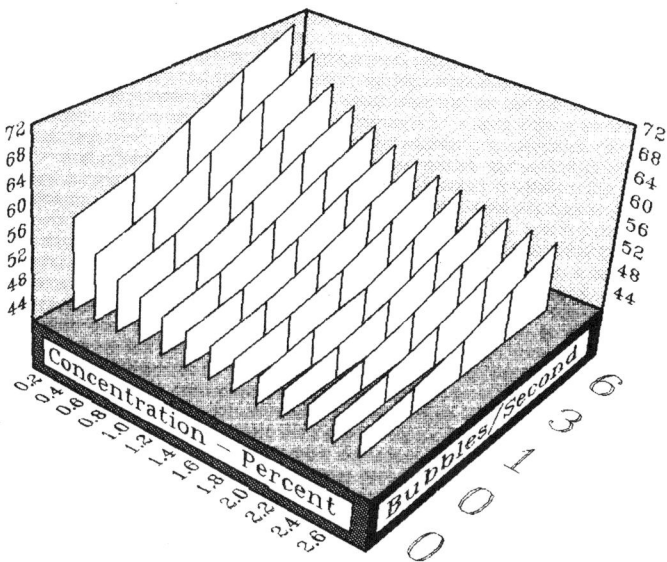

Figure 9. Surface Tension as a Function of Bubble Rate (Two Dimensional Test Data Plots at Various Concentrations).

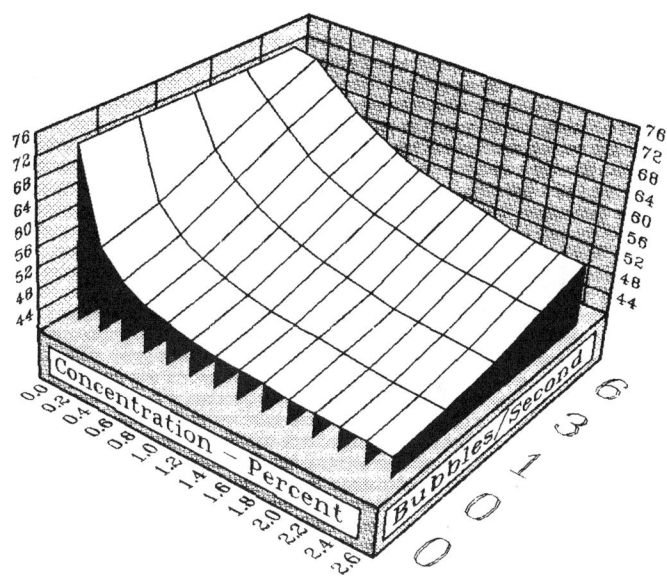

Figure 10. Dynamic Surface Tension Representation (Three Dimensional Characteristics).

Deionized water [surface tension of 72.9 dynes/cm. @ 20 degrees C.] and ethyl alcohol [surface tension of 22.4 dynes/cm, @ 20 degrees C.] were used as calibration standards.

PROCEDURE FOR DYNAMIC SURFACE TENSION DETERMINATION

In this method, the titration of the Regain NF was carried out in aqueous solution and monitored by using the SensaDyne® Model 6000 surface tensiometer. A series of five different bubble rates were selected and the SensaDyne instrument was calibrated at each bubble rate, prior to data collection.

The cell was placed on a laboratory stirrer, and the Regain NF was introduced into a 100 ml aqueous base in increments of 0.2 ml. After each incremental addition, the stirrer was turned on to mix the solution and then turned off to allow the solution to stabilize. When the surface tension reading stabilized after stirring, the data was captured and sent to a user-named ASCII file [R-1.TXT, etc.] by pressing a function key (F2) on the computer keyboard.

In addition to the file name, the user enters an alphabetical label [A to Z], and up to a thirty character comment. The resultant data is shown in Tables II through VI.

RESULTS AND DISCUSSION

Each of the five sets of data generated by the titration can be plotted in two dimensions, similar to Figure 2, as a function of surface tension versus concentration. Figure 7 shows two-dimensional test plots at the five bubble rate settings. The resultant two-dimensional graph shows a nest of five curves with the higher curve having the fastest bubble rates. The higher surface tensions over the same concentrations are the result of reduced dynamic surfactant migration times. These same curves can be better shown in a three-dimensional context by re-plotting them, as in Figure 8. Here, each two-dimensional curve is taken from Figure 7 and re-plotted along a third axis at corresponding bubble frequencies [bubbles/second].

The data sets in Tables II through VI can be further manipulated to form a series of two-dimensional plots of surface tension versus bubble frequency, with each plot at a different concentration, by plotting the data in all the "A' rows, "B" rows, etc. These plots are shown in Figure 9, in a three-dimensional context for clarity purposes.

The data in Tables II through VI are in ASCII format and can readily be imported into various spreadsheet or word processing programs for consolidation, and to develop the three-dimensional data matrix needed for a graphing program. Entering the data for surface tension, concentration, and bubble frequency into a three-dimensional graphing program

Table V. Experimental Data in ASCII Format for Run (R-4.TXT).

DATE	TIME	#	TEMP.	B/S	S.T.	COMMENTS
06-26-1992	09:51:32	A	21.4	2.97	67.8	0.2 ml Concentration
06-26-1992	09:55:36	B	21.4	3.02	63.8	0.4 ml Concentration
06-26-1992	10:03:14	C	21.4	3.34	60.3	0.6 ml Concentration
06-26-1992	10:05:32	D	21.4	3.43	57.9	0.8 ml Concentration
06-26-1992	10:09:37	E	21.4	3.55	56.0	1.0 ml Concentration
06-26-1992	10:13:52	F	21.4	3.63	54.4	1.2 ml Concentration
06-26-1992	10:16:29	G	21.4	3.63	53.5	1.4 ml Concentration
06-26-1992	10:18:37	H	21.4	3.76	52.3	1.6 ml Concentration
06-26-1992	10:21:33	I	21.5	3.80	51.0	1.8 ml Concentration
06-26-1992	10:24:42	J	21.5	3.92	50.5	2.0 ml Concentration
06-26-1992	11:27:18	K	21.5	4.08	50.1	2.2 ml Concentration
06-26-1992	11:29:07	L	21.5	4.14	49.5	2.4 ml Concentration
06-26-1992	11:30:52	M	21.5	4.24	49.2	2.6 ml Concentration

Table VI. Experimental Data in ASCII Format for Run (R-5.TXT).

DATE	TIME	#	TEMP.	B/S	S.T.	COMMENTS
06-26-1992	11:05:39	A	21.5	6.05	71.8	0.2 ml Concentration
06-26-1992	11:07:21	B	21.5	6.12	68.1	0.4 ml Concentration
06-26-1992	11:11:03	C	21.5	6.24	64.2	0.6 ml Concentration
06-26-1992	11:14:46	D	21.6	6.43	61.6	0.8 ml Concentration
06-26-1992	11:16:12	E	21.6	6.52	59.5	1.0 ml Concentration
06-26-1992	11:19:56	F	21.6	6.65	57.7	1.2 ml Concentration
06-26-1992	11:22:31	G	21.6	6.76	56.5	1.4 ml Concentration
06-26-1992	11:26:08	H	21.7	6.83	55.5	1.6 ml Concentration
06-26-1992	11:28:57	I	21.7	6.90	54.4	1.8 ml Concentration
06-26-1992	11:31:17	J	21.7	7.12	53.3	2.0 ml Concentration
06-26-1992	11:34:33	K	21.8	7.38	52.8	2.2 ml Concentration
06-26-1992	11:38:44	L	21.8	7.66	52.5	2.4 ml Concentration
06-26-1992	11:42:40	M	21.8	7.94	52.0	2.6 ml Concentration

results in the graph shown in Figure 10. This particular graph was done using "Foxgraph®" by Fox Software, although any similar three-dimensional plotting program can provide similar suitable results.

Figure 10 shows the effects on surface tension when concentration and bubble rates are varied. These results can be correlated to any time-dependent coating, spray, or dynamic fluid application process where a dynamic surfactant has a time constraint placed upon it due to the speed of the process. By testing the fluid or coating and obtaining the three-dimensional characteristics, operational and quality control problems can be more readily defined and mitigated.

The data can further be revised to reflect accurate interface development times instead of peak to peak bubble rates, as done here. More accurate interface development times would influence the data in Figures 8 through 10 such that the slopes at higher bubble rates would not be quite as pronounced as those shown. The overall three-dimensional graphs would be quite similar.

CONCLUSIONS

The surface tension data acquisition and three-dimensional graphic display method presented here has several advantages over other surfactant analysis techniques. By using a titration method, as described, a maximum amount of data can be accurately obtained in a minimum amount of time. The five data sets in Tables II through VI were collected over a period of less than four hours. This data represents the complete three-dimensional characterization of Regain NF. The method is kept quite simple by continuously monitoring surface tension during a single phase titration, and capturing only the values needed at points after each new solution was thoroughly mixed and stable.

Some limitations of this method can appear where coatings or other fluids are very viscous or have high solids percentages. Very high bubble rates may not be always achievable. Past work done with some very difficult polymers indicate that maximum rates of four or five bubbles per second are the best that can be obtained due to the much slower flow of the fluid. Bubble rate, to some degree depends on the ability and speed of the fluid to flow back into the void created by the bubble departing from the orifice. In some cases, the more limited amount of three-dimensional data obtainable may have to be further extrapolated into the area of interest.

The advantage of the method is very evident when data in Tables II through VI is compared with the single three-dimensional graph in Figure 10. The method can be applied, in general, to any fluid or coating that contains an active surfactant.

REFERENCES

1. Matijevic, E., ed., <u>Surface and Colloid Science</u>, Vol. **1**, Wiley-Interscience, (1969).

2. Rehfeld, S., "Adsorption of Sodium Dodecyl Sulfate at Various Hydrocarbon-Water Interfaces", J. Phys. Chem., **71(3)**, 738-745, (1967).

3. Sudgen, S., "The Determination of Surface Tension From the Maximum Pressure in Bubbles", J. Chem. Soc., **121**, 859-867, (1922).

4. Schork, F.J. and Ray, W.H., "On-line Monitoring of Emulsion Polymerization Reactor Dynamics", presented at ACS meeting, Las Vegas, (1980).

5. Jasper, J.J., "The Surface Tension of Pure Liquid Compounds", J. Phys. Chem. Ref. Data, Vol. **1(4)**, 841-859, (1972).

6. Weissberger, A. and Rossiter, B., Phys. Methods of Chemistry, Vol. **1**, Wiley-Interscience, (1959).

7. Gaines, G.L., Jr., "Surface and Interfacial Tension of Polymer Liquids-A Review", Polymer Eng. and Sci., 12(1), 1-11, (1972).

8. Heller, W. and Peters, J., "Mechanical and Surface Coagulation", J. Colloid Interface Sci., 32(4), 592-605, (1970).

9. Jho, C., "Effect of Pressure on the Surface Tension of Water: Adsorption of Hydrocarbon Gases and Carbon Dioxide on Water at Temperatures Between Zero and Five Degrees Centigrade", J. Colloid Interface Sci., **65(1)**, 141-154, (1978).

10. Woolfrey, S.G., Banzon, G.M. and Groves, M.J., "The Effect of Sodium Chloride on the Dynamic Surface Tension of Sodium Dodecyl Sulfate Solutions", J. Colloid Interface Sci., **112(2)**, 583-587, (1986).

11. Bassemir, R.W. and Krishnan, R., "Practical Applications of Surface Energy Measurements in Flexography", <u>Flexo</u>, pp. 31-40, July (1990).

12. Dougherty, W.R., "Acetylenic Diol Surfactant Cut Foaming and Wetting Problems", Adhesives Age, September, 26-30 (1989).

13. Schramm, L.L., Smith, R.G. and Stone, J.A., "A Surface Tension Method for the Determination of Anionic Surfactant in Hot Water Processing of Athabasca Oil Sands", Colloids and Surfaces, **11**, 247-263, (1984).

14. Lusk, L.T., Cronan, C.L., Chicoye, E., and Goldstein, H., "A Surface-Active Fraction Isolated from Beer", presented at the American Society of Brewing Chemists, Inc. 50th Annual Meeting, St. Louis, pp. 91-95, (1984).

SURFACTANTS IN COATINGS AND PAINT

APPLICATIONS

Ashwin V. Parikh

Research and Development
Oleochemicals/Surfactants Group
Witco Corporation
3200 Brookfield Street
Houston, TX 77045

ABSTRACT

Several properties of the surfactants and their applications in waterborne coatings and paint formulations were discussed. These formulations are in colloidal form, which contained polymer as a binder, pigments, solvents, water and surfactants as stabilizers, emulsifiers, wetting agents, dispersants, defoamers etc. in order to obtain the desired performance properties of the coated films. The selection and performance of a given surfactant depend on the structure and functional groups. Alcohol ethoxylates, sulfosuccinates and alkanolamides are extensively used as emulsifying agents in latex formation and paint stabilization to reduce particle size in order to prevent settling of the dispersed phase. A method of paint evaluation and performance characterization was described in order to determine proper surfactant selection for optimum properties of the coating and paint films.

INTRODUCTION

Emulsion paints, in general, are obtained by mixing a polymer latex, pigments, surfactants, solvents, water and other ingredients which provide unique properties to the

Surface Phenomena and Latexes in Waterborne Coatings and Printing Technologies, Edited by M.K. Sharma, Plenum Press, New York, 1995

203

formulated paint. The surfactants are used as additives in coating and paint formulations are depicted in Table I. The Polymer Latex is synthesized by emulsion polymerization of a monomer or monomers in aqueous solution containing suitable surfactants and initiator[1-5]. The widely used polymers include: acrylic polymers, polyesters, polyurethane, cellulosic polymers and polyamides. Among these polymers, waterborne acrylic polymers are recently developed for coatings and paint formulations by emulsion polymerization process. The emulsion polymerization process can allow to modify a latex polymer composition and morphology during latex synthesis. The surfactant type and concentration can significantly influence the latex particle size and size distribution, dispersion stability and morphology of the latex particles[6-10], which in turn can affect the coatings, printing inks, paint performance.

Apart from the surfactants in the polymer latex, a paint formulation requires additional surfactants in order to impart stability and special properties to the paint. This paper reports the basic selection criteria of such surfactants for the paint formulation. An overview of preparation of paints and their evaluation is also presented.

Table I. Different uses of Surfactants in Coatings and Paints.

DISPERSANTS (SEPARATORS)	WETTING AGENTS (STABILIZERS)	DEFOAMERS (FLOW CONTROLERS)
Polycarboxylates Polyacrylates Naphthalene-sulfonates	Alcohol Ethoxylates Sulfosuccinates Alkanolamides	Paraffins Metal Soaps Silicones Fattyacidesters Polysiloxanes
*Permanent Particle Separation *No Reduction in Particle Size	*Emulsifies *Reduces Particle Size and Surface Tension *Prevents Settling	*Defoamers Destroys Foam *Antifoam Prevents Foam Formation

EXPERIMENTAL

The emulsion paints can be formulated by mixing desired additives in the latex to obtain the optimum end-use properties of the coating and paint films. The ingredients required for the formulation of the latex paint are recorded in Table II.

The process of the latex paint formulation can be divided in two parts. The first phase is called the grind phase, where most of the ingredients are added to pigment and ground into a flowing liquid having a desired particle size (usually a 5 hegman grind). This process involves high speed agitation. In the second phase, known as the "Let Down Phase", the latex binder and a few other additives with water is added to the pigment grind and mixed under a slow speed mixer. Defoamers are added in both phases to prevent formation of foam or to destroy the foam. The defoamer added to prevent foam at pigment grind stage is generally known as antifoaming agent. The generic contents and their amount incorporated in emulsion paint formulations are listed in Table III.

The selection of these ingredients determines the end-use properties of the coatings and paint films on a given substrate. The paint can be formulated for an exterior or interior applications with high gloss, semi-gloss and low-gloss (e.g. flat paint).

Table II. A List of Ingredients Used in the Latex Paint Formulations.

* WATER	* DISPERSANTS
* DEFOAMERS	* WETTING AGENTS
* GLYCOLS	* PRESERVATIVES/ MILDEWCIDES
* COALESCING AGENTS	* EXTENDERS
* PRIME HIDING PIGMENTS	

HIGH SPEED AGITATION...... GRIND PHASE

```
LATEX BINDER
DEFOAMER
ADDITIVES
WATER/THICKNER
```

SLOW SPEED MIXING....... LET DOWN PHASE

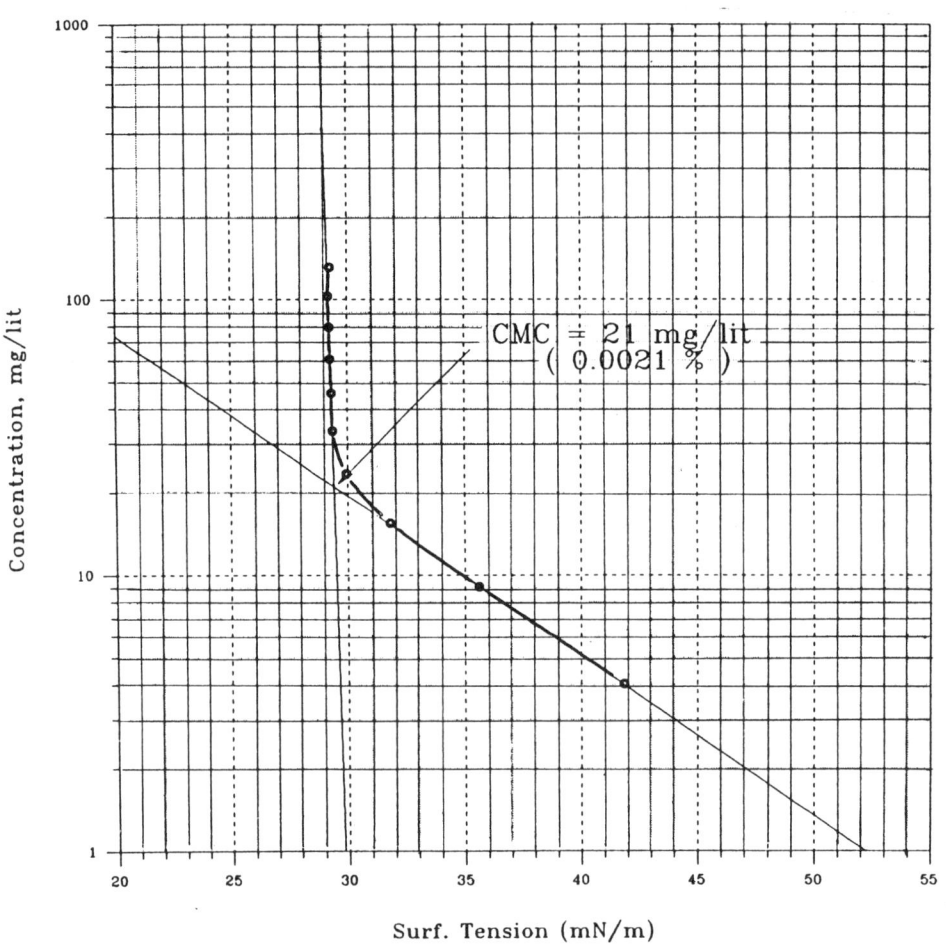

Figure 1. Surface Tension as a Function of Concentration for Witflow-60 Surfactant.

Table III. A Typical Composition of Emulsion Paint[11].

INGREDIENTS	AMOUNT (wt%)
Opaque Pigment	20.0
Extender Pigment	15.0
Pigment Dispersant	0.1
Protective Colloid	1.2
Latex	40.0
Preservative	0.5
Fungicide	optional
Coalescing Agent	2.0
Defoamer	0.1
Thickener	0.5
Water	20.6

RESULTS AND DISCUSSION

SURFACTANT PROPERTIES AND SELECTION

Surfactants used in paint formulation function as dispersants, wetting agents and defoamers. These properties arise from the unique functionalities on the surfactant molecule. For example, surfactants containing polycarboxylate groups act as dispersants. Table I lists the different properties of the surfactants and the functional groups employed in achieving the desired property of the surfactant. Defoamers function either by preventing the foam formation or destroying the foam already formed.

Surfactants are selected on the basis of the following criteria:

1. Functionality
2. Compatibility
3. Cost Effectiveness
4. Availability

The importance of the functionality was already discussed above. Compatibility of the surfactant in the paint formulation is also important because, wrong selection of surfactant can destabilize the polymeric latex and lead to instability of the paint itself. For example, a cationic surfactant added to a paint formulation, containing polymeric latex stabilized by anionic surfactant will lead to coagulation of the paint. Cost effectiveness can be achieved by selecting a few surfactants to perform several functions or effective surfactants which will bring about maximum performance with minimum dosage level. A list of Witco Surfactants, defining the structure and functionality is presented in addendum.

SURFACE TENSION AND CMC OF SURFACTANTS

A surfactant possesses the property of absorbing onto the surfaces or interfaces of a system and alters the free energy of the interface. The interfacial tension between two phases is measured in terms of interfacial free energy per unit area of the interface. This gives the interfacial tension. The interfacial free energy between the liquid and air is the surface tension of the liquid. Surfactants reduce the interfacial free energy. The surface tension of water is reduced by the presence of surface active material dissolved in water. The decrease in surface tension is dependent on the amount of surfactant in solution. A plot of surface tension against concentrations of surfactant (Witflow-60) is shown in Figure 1. The sharp break in the curve occurs at a specific concentration of surfactant, which is called the Critical Micelle Concentration (CMC). The critical micelle concentration is the minimum concentration of surfactant needed to produce micelles or colloidal particles of surfactants in solution. The dosage level of surfactants for proper dispersion of pigments in paints is determined on the basis of CMC. Normally surfactant concentration above the CMC is recommended for paint formulation.

EMULSION PAINTS

The major ingredients of emulsion paints, as mentioned earlier, are polymeric latex and pigments. There are several other ingredients added to the paint formulation. Among the ingredients listed in Table II, the defoamers are one of the important ingredients in a paint formulation. These defoamers are added in two stages in a paint formulation. Emulsion paints are used in interior and exterior applications.

Table IV. Composition and Properties of Semi-Gloss Paint.

```
DENSITY           :  10.49
SPECIFIC GRAVITY :  1.26
P.V.C.            :  22.26 %
P/B RATIO         :  1.029
SPREAD AT 1 MIL  :  538 SQ.FT.
COATING VOC       :  226 G/L
MAT'L VOC         :   97 G/L
```

	WEIGHT	VOLUME
TOTAL VEHICLE	76.13 %	92.53 %
PIGMENT	23.87 %	7.47 %
VOLATILE	52.94 %	66.43 %
ORG.SOLVENT	7.75 %	9.56 %
NON-VOLATILE	47.06 %	33.57 %
NON-VOL VEH	23.20 %	26.10 %

Table V. A Typical Composition of Formulated Interior Flat Paint.

MATERIAL	LB/100 U.S.GALLONS	U.S. GALLONS	% BY WEIGHT
GRIND			
WATER	420.29	50.45	35.96
THICKENER-HEC	5.51	0.48	0.47
PRESERVATIVE	2.30	0.27	0.20
ANIONIC DISP.	4.61	0.50	0.39
KTPP	1.00	0.13	0.09
NONIONIC DISP.	2.20	0.25	0.19
AMP-95	1.00	0.13	0.09
PROPYLENE GLYCOL	17.34	2.01	1.48
BUBBLE BREAK.748	1.90	0.26	0.16
TITANIUM DIOXIDE	150.32	4.37	12.86
CLAY	125.27	5.82	10.72
CALCIUM CARB.	200.42	8.87	17.15
LET DOWN			
VA/BA LATEX	226.78	25.20	19.40
FILMING AGENT	7.92	1.00	0.68
BUBBLE BREAK.748	1.90	0.26	0.16
TOTAL	1168.75	100.00	100.00

Table VI. The Properties of Formulated Flat Paint.

```
DENSITY           : 11.69
SPECIFIC GRAVITY :  1.40
P.V.C.            : 58.53 %
P/B RATIO         :  3.632
SPREAD AT 1 MIL  :  539 SQ.FT.
COATING VOC       :   93 G/L
MAT'L VOC         :   35 G/L
```

	WEIGHT	VOLUME
TOTAL VEHICLE	58.72 %	80.33 %
PIGMENT	41.28 %	19.67 %
VOLATILE	47.35 %	66.40 %
ORG.SOLVENT	2.46 %	3.43 %
NON-VOLATILE	52.65 %	33.60 %
NON-VOL VEH	11.37 %	13.94 %

Depending upon the pigment volume concentration in the paint, these paints can be classified into Semi-gloss or Flat. The composition and properties of the semi-gloss paints are shown in Table IV , while composition and properties of the flat paint are listed in Tables V and VI.

The properties of the paint are evaluated by performing several tests. A collection of tests and reporting of test results are included in the addendum. A systematic evaluation of the paint can be completed by using these parameters.

REFERENCES

1. Piirma, I., Editor, "Emulsion Polymerization" Academic Press, New York, NY, (1982).

2. Daniels, E.S., Sudol, E.D. and El-Aasser, M.S., Editors, "Polymer Latexes: Preparation, Characterization and Applications", ACS Symposium Series No. 492, (1992).

3. Athey, R.J., Jr., "Emulsion Polymer Technology" Marcel Dekker, New York, NY, (1991).

4. Tang, P.L., Sudol, E.D., Adams, M.E., Silebi, C.A. and El-Aasser, M.S., Miniemulsion Polymerization: In "Polymer Latexes: Preparation, Characterization and Applications", (Daniels, E.S., Sudol, E.D. and El-Aasser, M.S., Editors), ACS Symposium Series No. 492, 72,(1992).

5. Lee, K.C., El-Aasser, M.S. and Vanderholff, J.W., Batch and semicontinuous emulsion copolymerization of vinylidene chloride and butyl methacrylate I: Kinetics of VDC-BMA emulsion polymerization and surface and colloid properties of VDC-BMA latexes, J. App. Polym. Sci., **45**, 2207, (1992).

6. Urquiola, M.B., Sudol, E.D., Dimonie, V.L. and El-Aasser, M.S., Emulsion polymerization of vinyl acetate using a polymerizable surfactant. III: Mathematical model, J. Polym. Sci.: Part-A: Polym. Chem., **31**, 1403-1415, (1993).

7. Shen, S., Sudol, E.D. and El-Aasser, M.S., Control of particle size in dispersion polymerization of methyl methacrylate, J. Polym. Sci.: Part-A: Polym. Chem., **31**, 1393-1402, (1993).

8. Urquiola, M.B., Arzamendi, G., Leiza, J.R., Zamora, A., Asua, J.M., Delgado, J. El-Aasser, M.S. and Vanderholff, J.W., Semicontinuous seeded emulsion polymerization by vinyl acetate and methyl acrylate, J. Polym. Sci., Part-A: Polym. Chem., **29**, 169, (1991).

9. DosRamos, J.G. and Silebi, C.A., The determination of particle size distribution of submicrometer particles by capillary hydrodynamic fractionation (CHDF), J. Colloid Interface Sci., **135(1)**, 165-177, (1990).

10. Durali, M., An investigation into the structure and breakup of aggregated latex particles, Ph. D. Dissertation, Lehigh University, Bethlehem, PA., pp.215, (1993).

11. Martins, C.R., Waterborne Coatings: Emulsions and Water- Soluble Paints, Van Nostrand Reinhold Company, p.50, (1981).

ADDENDUM

```
                    ┌─────────────────────────────┐
                    │                             │
                    │    WITCO SURFACTANTS        │
                    │                             │
                    └─────────────────────────────┘
```

                                     O
                                     ‖
ETHOXLATED FATTY   R−C−O(CH₂CH₂O)ₙH   R = fatty radical
           ACIDS

                                     O
                                     ‖
PROPOXYLATED FATTY   R−C−O(CH₂CHO)ₙH   R = fatty radical
            ACIDS   │
                                     CH₃

SULFONATED FATTY   CH₃−CH₂(CH₂)ₙ−COONa
          ACIDS   │
                               SO₃Na

                                        OH
                                        │
POLYGLYCEROL ESTERS   R−CO(CH₂CHCH₂O)ₙH   R = fatty radical
                           ‖
                           O

                              O
                              ‖
                        R−COCH₂
                             │
                           HCOH
                             │
                           H₂COH
FATTY MONO- AND   R = fatty radical
DIGYLCERIDES

                              O
                              ‖
                        R−COCH₂
                             │
                           HCOH
                             │
                           H₂COC−R
                                ‖
                                O

ALCOHOL SULFATES $\qquad$ $CH_3(CH_2)_n CH_2O-SO_3X$ $\qquad$ X = ammonium, sodium or amine

ALCOHOL ETHER SULFATES $\qquad$ $CH_3(CH_2)_n O (CH_2CH_2O)_n - SO_3X$

ETHOXYLATED FATTY ALCOHOLS $\qquad$ $RO-(CH_2CH_2O)_n H$ $\qquad$ R = fatty alcohol

PROPOXYLATED FATTY ALCOHOLS $\qquad$ $RO-(CHCH_2O)_n H$ $\qquad$ R = fatty alcohol

with CH_3

ETHOXYLATED, PROPOXYLATED ALCOHOLS $\qquad$ $RO-(CH_2CH_2O)_n (CHCH_2O)_m H$ $\qquad$ R = fatty alcohol

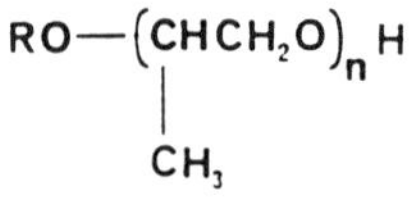

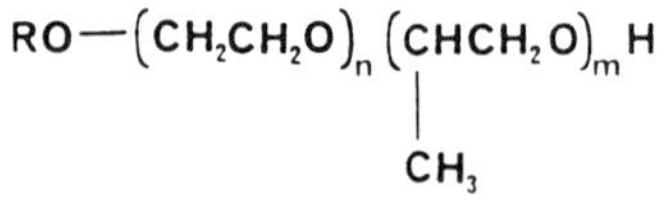

with CH_3

ALKANOLAMIDES (DEA) $\qquad$ 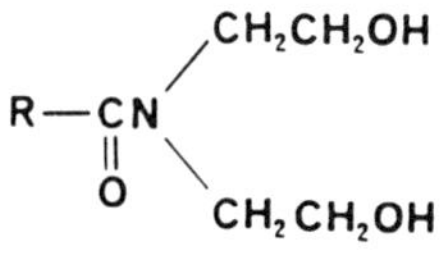
$$R-C\underset{\underset{O}{\parallel}}{N}\begin{cases} CH_2CH_2OH \\ CH_2CH_2OH \end{cases}$$
R - coconut, lauryl myristyl, stearyl oleyl radical

ALKANOLAMIDES (MEA) $\qquad$ 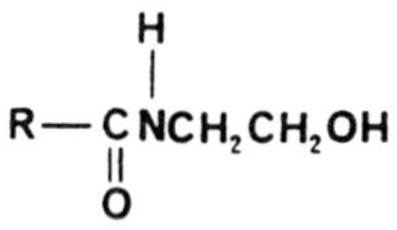
$$R-\underset{\underset{O}{\parallel}}{C}\overset{H}{N}CH_2CH_2OH$$
R - coconut, lauryl myristyl, stearyl oleyl radical

ALKANE SULFONATES $\qquad$ $CH_3-(CH_2)_n-CH_2-CH_2-SO_3Na$

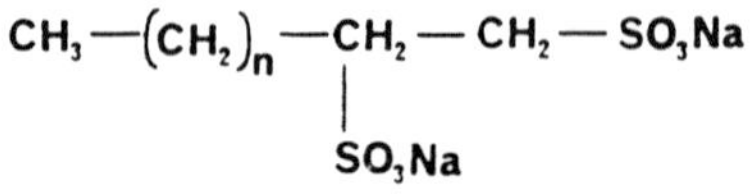

$CH_3-(CH_2)_n-CH_2-CH_2-SO_3Na$

with SO_3Na

Mixture of n-alkane mono and di-sulfonates

216

SODIUM ISETHIONATE $\quad$ HO—CH$_2$CH$_2$—SO$_3$Na

ETHOXYLATED ALKYL PHENOLS

R—⬡—O(CH$_2$ CH$_2$O)$_n$H

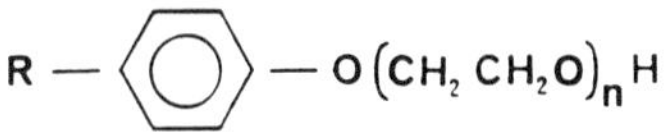

R = octyl, nonyl, decyl, dodecyl etc.

ETHOXYLATED ALKYL PHENOL SULFATES

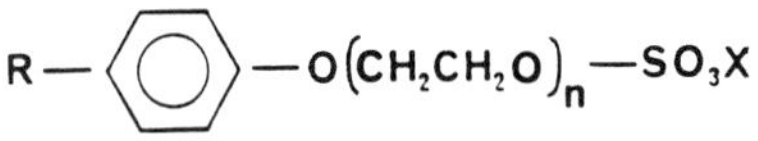

R—⬡—O(CH$_2$CH$_2$O)$_n$—SO$_3$X

x = ammonium, sodium or amine

ETHOXYLATED FATTY AMINES

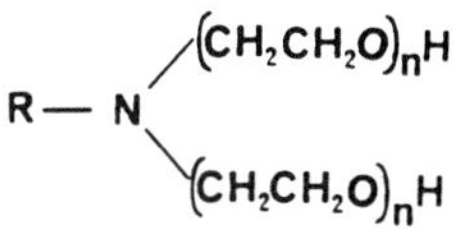

$$R—N \begin{cases} (CH_2CH_2O)_nH \\ (CH_2CH_2O)_nH \end{cases}$$

R = fatty radical

AMINES OXIDES

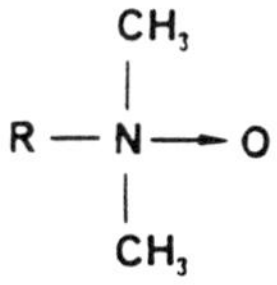

$$R—\overset{\displaystyle CH_3}{\underset{\displaystyle CH_3}{N}}→O$$

R = coconut fatty radical

ALPHA OLEFIN SULFONATE

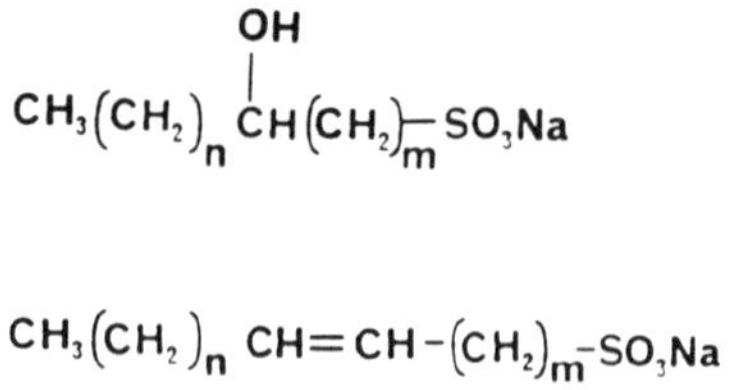

$$CH_3(CH_2)_n \overset{\displaystyle OH}{\underset{}{CH}}(CH_2)_{\overline{m}}SO_3Na$$

Blend of hydroxy alkane and alkene sulfonates

$$CH_3(CH_2)_n\ CH=CH-(CH_2)_{\overline{m}}SO_3Na$$

m = 2 - 4
n = 8 -12

COCOAMIDOPROPYL BETAINE

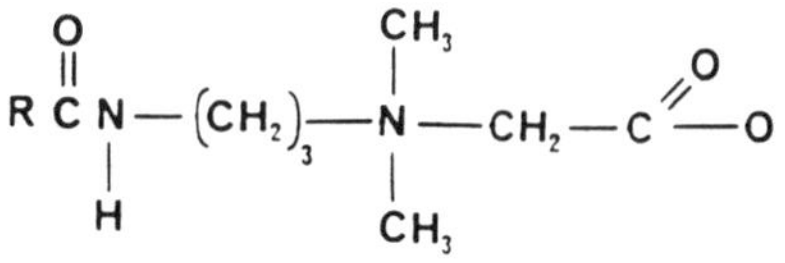

$$R\overset{\displaystyle O}{\underset{\displaystyle H}{C}}N-(CH_2)_3-\overset{\displaystyle CH_3}{\underset{\displaystyle CH_3}{N}}-CH_2-C\overset{\displaystyle O}{\diagdown}O$$

R = coco group

CARBOXYLATED NONIONIC

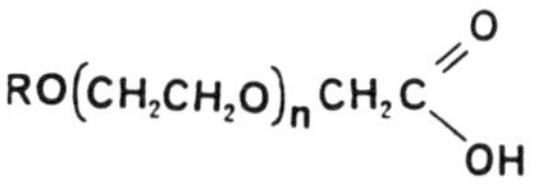

$$RO(CH_2CH_2O)_n CH_2C\overset{\displaystyle O}{\underset{\displaystyle OH}{\diagup}}$$

R = nonyl phenol, fatty alcohols, alcohol ethoxylates

GLYCEROL ETHOXYLATES

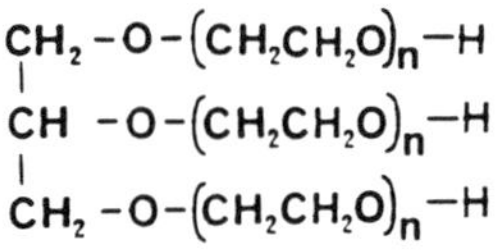

$$CH_2-O-(CH_2CH_2O)_n-H$$
$$CH\ -O-(CH_2CH_2O)_n-H$$
$$CH_2-O-(CH_2CH_2O)_n-H$$

POLYETHYLENE GLYCOL ESTERS

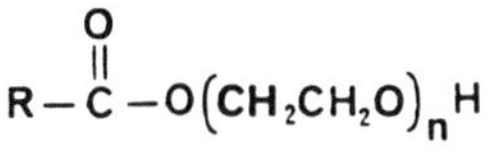

$$R-\overset{\displaystyle O}{C}-O(CH_2CH_2O)_nH$$

R = fatty radical

IMIDAZOLINES

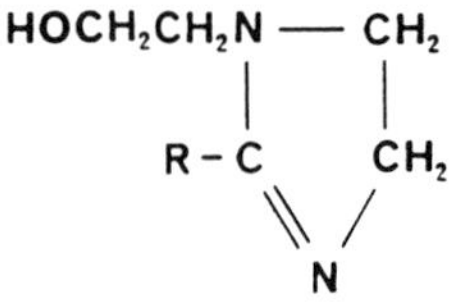

$$HOCH_2CH_2N-CH_2$$
$$R-C \quad CH_2$$
$$\diagdown N \diagup$$

R = tall oil fatty radical

ALKYL NAPHATHALENE SULFONATES

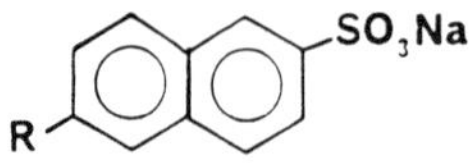

R — (naphthalene) — SO₃Na

NAPHTHALENE SUL-FONATE FORMALDEHYDE CONDENSATES

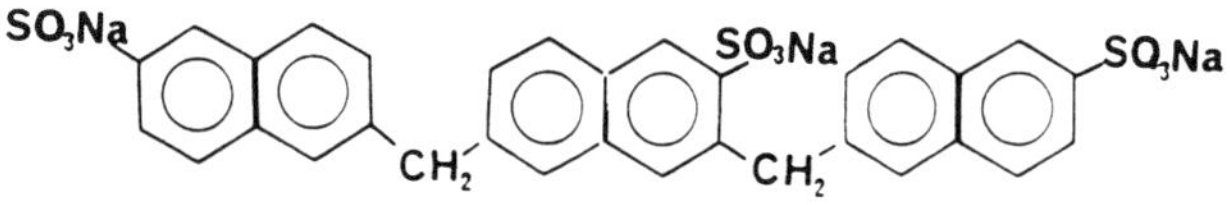

SO_3Na SO_3Na SO_3Na

CH_2 CH_2

SORBITAN ESTERS

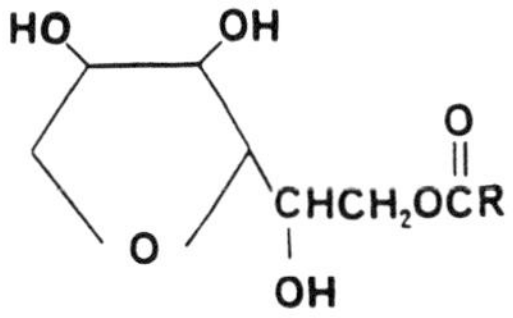

HO   OH

$CHCH_2OCR$ O

OH

O

R = fatty alkyl group

SORBITAN ESTERS ETHOXYLATES

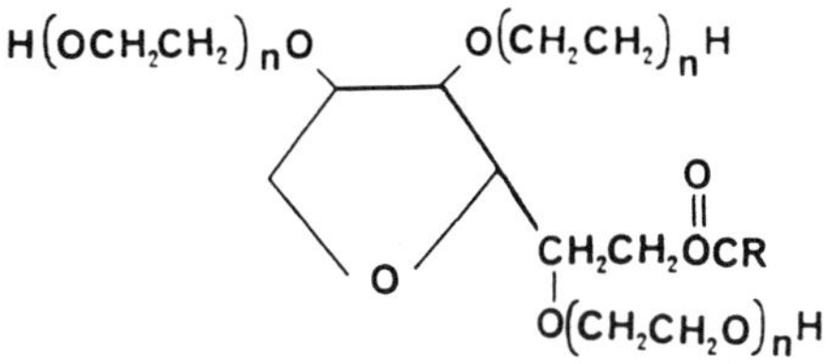

$H(OCH_2CH_2)_nO$ $O(CH_2CH_2)_nH$

CH_2CH_2OCR O

$O(CH_2CH_2O)_nH$

O

R = fatty alkyl group
N = moles of ethylene oxide

ALKYLBENZENE SULFONATES

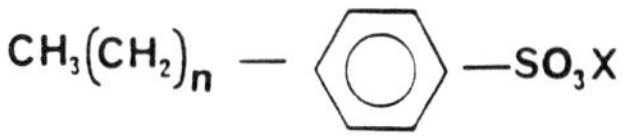

$CH_3(CH_2)_n$ —⟨◯⟩— SO_3X

X = sodium, calcium, amine, etc.

SODIUM XYLENE SULFONATE

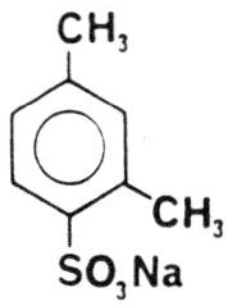

CH_3

CH_3

SO_3Na

SULFOSUCCINATES
(mono-, di-, and unsymetrical esters)

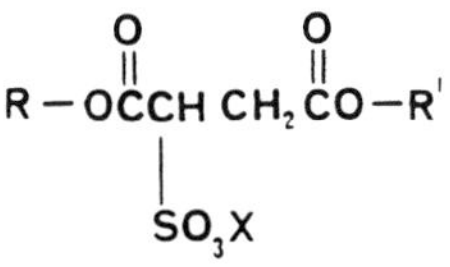

$R-OCCH\,CH_2\,CO-R'$ O   O

SO_3X

R = alcohol, alcohol ethoxylate,
X = sodium, amine, etc.

PHOSPHATE ESTERS
(mono/diesters)

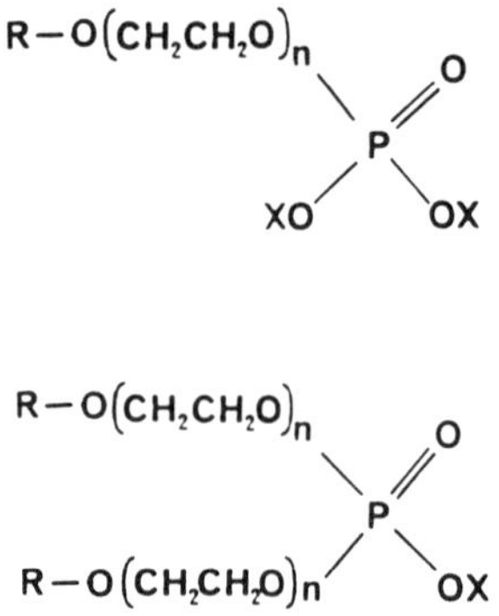

$$R-O(CH_2CH_2O)_n \quad \underset{XO}{\overset{O}{\underset{|}{P}}} \quad OX$$

$$R-O(CH_2CH_2O)_n$$
$$R-O(CH_2CH_2O)_n \quad \underset{}{\overset{O}{\underset{|}{P}}} \quad OX$$

R = alcohol,
    alcohol ethoxylate,
    alkyl phenol,
    alkyl phenol
        ethoxylate

POLYPROPOXY QUATERNARY AMMONIUM

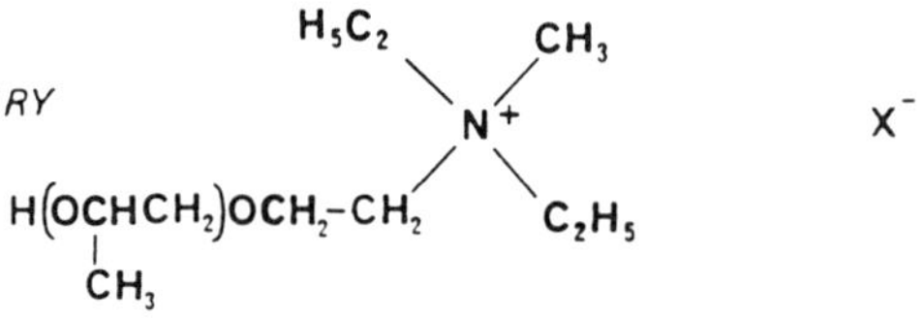

$$H(OCHCH_2)OCH_2-CH_2 \overset{H_5C_2}{\underset{CH_3}{\overset{|}{N^+}}} \overset{CH_3}{\underset{C_2H_5}{}}$$

X^-

X^- = phosphate
    = chloride or
    acetate ion

ALKYLDIMETHYLBENZYL AMMONIUM CHLORIDE

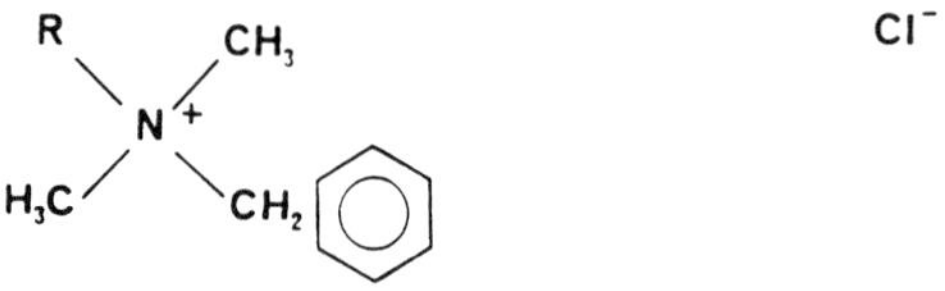

$$\overset{R}{\underset{H_3C}{\overset{|}{N^+}}}\overset{CH_3}{\underset{CH_2}{}}$$

Cl^-

R = lauryl or
    stearyl radical

N(STEAROYL/LAUROYL) COLAMINOFORMYLMETHYL PYRIDINIUM CHLORIDE

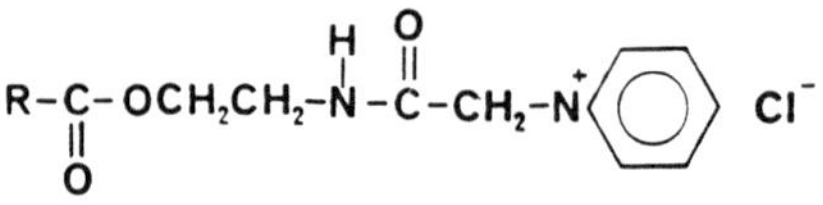

$$R-\overset{O}{\underset{||}{C}}-OCH_2CH_2-\overset{H}{\underset{}{N}}-\overset{O}{\underset{||}{C}}-CH_2-N^+ \quad Cl^-$$

```
#1      HEGMAN GRIND

#2      VISCOSITY @_____° C

        FRESH
        OVERNIGHT

#3      pH @_____° C
        FRESH
        OVERNIGHT

#4      HEAT STABILITY @ 49°C (120°F)
        ONE WEEK: KU
                  pH
           6 mils  D/D

#5      FREEZE-THAW
                    CYCLE 1
                    CYCLE 2
                    CYCLE 3 KU
                            pH
                    6 mils  D/D
                    CYCLE 4
                    CYCLE 5 KU
                            pH
                    6 mils  D/D

#6      OVERNIGHT 6 mils  D/D

        FORM 3B    60° gloss
        OPACITY    85° sheen

        FORM 1B
        PENOPAC

        FORM 7B
        SAG & LEVELING

#7      4.4°C (40°F)
        COALESCENCE

#8      BRUSHABILITY

        LAPPING 5, 10, 20 minutes

        TOUCHUP "X"

#9      COLOR DEVELOPMENT
        (150g + 1.5g) 5 min.

#10     COLOR ACCEPTANCE
        FINGER RUBUP

#11     ADHESION %
        CRS-Q PANEL
```

#12   QUV EXPOSED____HRS.

       60° gloss BEFORE
                 AFTER

       85° sheen BEFORE
                 AFTER

#13   SCRUB RESISTANCE

       START WEAR CYCLES:
       25%   "   "   :
       50%   "   "   :
       Stop at   "   :

#14   EXTERIOR TEST FENCE EXPOSURE:

       MOLD: : North - Under Eve/Offset:

       DURABILITY & WEATHERING: South/Offset Verticle:

       CORROSION   : 45° South/Offset North

#15   EMULSION TYPE:

#16   FORMULATION NO:

       DESCRIPTION

MULTI-PANEL PAINT INSPECTION SHEET

TEST NO. _____ PROBLEM NO. _____ COLOR _____ INSPECTED BY _____

LOCATION _____ EXPOSURE _____ N S E W VERT. 45° _____ DATE _____

PANEL OR PAINT NO.	PROPERTIES	REMARKS

SINGLE PANEL PAINT RECORD FORM

TEST NO._____ PROBLEM NO. _____ PAINT NO. _____

SUBSTRATE _____ LOCATION _____ COLOR _____

PURPOSE OF TEST_____

EXPOSED_____ REMOVED_____ EXPOSURE_____ N S E W____ VERT.___45°

EXPOSURE TIME   X GENERAL APPEARANCE

3 6 9 12 15 18 21 24 27 30 33 36 39 42 45 48 51 54 57 60

10 8 6 4 2 0

X GLOSS   O CHALKING   ● EROSION

X CHECKING   O CRACKING

X FLAKING   O SCALING   ● PEELING

X DIRT   O MILDEW   ● RUSTING

X FADING   O DARKENING   ● YELLOWING

RATING

INSPECTED BY _____

COMPOSITION

I ST COAT	2 ND COAT	3 RD COAT	4 TH COAT

REDUCTION

I ST COAT	2 ND COAT	3 RD COAT	4 TH COAT

MISCELLANEOUS

	I ST COAT	2 ND COAT	3 RD COAT	4 TH COAT
WEIGHT PER GALLON				
CONSISTENCY				
HIDING POWER				
PIGMENT VOLUME				
SPREADING RATE				
WORKING QUALITIES				
LEVELING				
DRYING CHARACTERISTICS				
APPLICATION CONDITIONS				
DRYING CONDITIONS				
ATMOSPHERIC CONDITIONS				
CONDITION OF THE SURFACE				
PREPARATION OF THE SURFACE				
PROTECTION OF THE BACK				

REMARKS:

AUTHOR INDEX

SUBJECT INDEX

Untreated (continued)
  polyethylene,131,133,134,
    136,137

Variance,21,27
Viscoelastic,91,93,95-97,
    99-102
Viscosity,1-4,6,7,9,11-17,20,
    24,27,34,39,41,43-45,47-
    53,59,68,140,146,175,176,
    178-180,182,185,221
Visible light,21
Voids,25

Water
  dispersible,42-44,58,60
  resistance,42,44,66,68
  sensitive,42
  soluble,42-44,175,182,185
Wavelength,21,143
Web-offset,103
Wet adhesion,115-121
Wettability,25,163
Wetting,3,25,31,58,81,83
    84,123-125,130,131,
    133,136

Wetting agent,25,64,84,
    203-205,207
White pigment,21
Whiteness,23,31,35,36
Whiteness index,31,35,36,
    38

Xenon
  atom,168
  gas,168
  NMR,167,171,172
  shift,168
X-ray diffraction,85

Yield,37,118
Yield stress,178

Zero point of charge
(ZPC),161,163,165
Zeta potential,84,85,
    88,153,154,156-165
Zone
  dynamic pumping,141
  shear-attrition,140
  shearing,140
  static,188